MW00466929

UNCERTAINTY

UNCERTAINTY

How It Makes Science Advance

Kostas Kampourakis
UNIVERSITY OF GENEVA

Kevin McCain
UNIVERSITY OF ALABAMA
AT BIRMINGHAM

OXFORD
UNIVERSITY PRESS

OXFORD
UNIVERSITY PRESS

Oxford University Press is a department of the University of Oxford. It furthers
the University's objective of excellence in research, scholarship, and education
by publishing worldwide. Oxford is a registered trade mark of Oxford University
Press in the UK and certain other countries.

Published in the United States of America by Oxford University Press
198 Madison Avenue, New York, NY 10016, United States of America.

© Oxford University Press 2020

CIP data is on file at the Library of Congress
ISBN 978-0-19-087166-6

9 8 7 6 5 4 3 2 1

Printed by Sheridan Books, Inc., United States of America

For our children, Kaison James McCain, Giorgos Kampourakis, and Mirka Kampouraki; may you come to understand uncertainty and have no fear of it.

CONTENTS

PART III **Accepting Uncertainty in Science**

PREFACE

"Science." This term is used to refer to a variety of theories, models, explanations, concepts, practices, attitudes, approaches, and more that relate to the study of the natural world. Philosophers have shown that there is no single entity to refer to as science; instead, there are a variety of disciplines that study the natural world with a variety of methods and explanatory aims. Therefore, it is more appropriate to refer to "sciences," plural, in order to account for this variety. However, it seems that in general we are stuck with the term "science," especially when it comes to the public sphere. Even though lay people may not be able to provide a concise definition of what "science" is, they most likely have no doubt about what the term means. This becomes evident in polls that aim at documenting their views about and attitudes toward "science." Therefore, we have decided to use the term "science" throughout this book to refer to the habits of mind, methods, approaches, and body of knowledge that aim to provide an empirical understanding of the natural world.

Inspired by Stuart Firestein's books *Ignorance* and *Failure*, we decided to write a book that discusses another important and inherent feature of science: uncertainty. Our goal is to show that uncertainty is an inescapable feature of science that, counterintuitive as it might sound, does not prevent us from gaining scientific knowledge and understanding. As individuals, and as societies, we need to understand and accept uncertainty in science. Perhaps surprisingly, it is because of this feature that science has made, and still makes, important advancements. Uncertainty motivates further research and a better understanding of natural phenomena. Scientific knowledge is perhaps the sort of knowledge that is as close to certainty as we can get, but this only means that it is closer to being certain than other kinds of knowledge, not that it is absolutely so. Whereas nothing compares to scientific knowledge when it comes to understanding the natural world around us, such knowledge is at the same time flexible enough to accommodate new findings. This happens because continuous research deals better and better with uncertainty and produces an increasingly reliable body of knowledge on the basis of solid evidence and rational thinking. Uncertainty is thus the driving force in this quest: it leads to more research, which in turn serves the ultimate goal of science—understanding.

People have strived for certainty throughout the centuries through oracles, horoscopes, and other superstitious endeavors. However, the only certainty in human life is death, and even when this will occur is highly uncertain. At times people have tried to remove uncertainty from their lives by adopting various forms of fundamentalism (religious, ideological, etc.) that provide them with a false sense of certainty that often has dangerous consequences. In recent years, some people have turned to science in the quest for certainty. It is unfortunate that this at times leads to a sort of fundamentalism as well, which arises when people consider (and expect) science to be certain. But this is a misunderstanding of the nature of science. Even though we can be extremely confident about the conclusions of science, it is full of uncertainties that we need to understand and appreciate.

Our aim is to reach a broad audience and provide them with an authentic account of how science is done, what difficulties scientists face, the knowledge and understanding they can arrive at, and why uncertainty is an inherent feature of all these. We believe that people need to understand that uncertainty in science is not a problem; being unaware of uncertainty and reaching unfounded conclusions is the real problem. In contrast, awareness of the uncertainties inherent in all scientific practices may allow for a more solid and deeper understanding of science and the natural phenomena it studies. This is how uncertainty actually makes science advance. We hope that by the end of the book the reader will have understood and appreciated the impact of uncertainty in science. We must note at this point that our coverage is far from exhaustive. In order to maintain a reasonable length, we have only touched on the important topics and limited our case studies to just six scientific domains. The interested reader who wants to explore some of the topics of this book further will find several excellent books and articles mentioned in the Notes at the end of the book.

We are indebted to Joan Bossert, our editor at Oxford University Press, who supported this project from the start and steered it in the right direction. As we have mentioned, the inspiration for writing this book came when we read Stuart Firestein's books *Ignorance* and *Failure,* and we thought that a similar book on uncertainty in science might be as useful as these. Incidentally, we are extremely happy that, like the Firestein books, this one is published by Oxford University Press. We are also indebted to the following scholars for reading the manuscript and kindly providing us with feedback: Gerd Gigerenzer, Sheldon Krimsky, Molly McCain, Ted Poston, Henk de Regt, and Michael Ruse. Finally, we are grateful to Phil Velinov as well as the staff at Newgen who worked with us toward the publication of the present book.

This book is dedicated to our children with the hope that they will come to understand uncertainty and have no fear of it. We hope that the same is true of you after you finish reading it.

PART I

Dealing with Uncertainty

1 | Uncertainty in Everyday Life

Our new Constitution is now established, and has an appearance that promises permanency; but in this world nothing can be said to be certain, except death and taxes.

—Benjamin Franklin, in a letter to Jean-Baptiste Leroy, 1789[1]

What Is Knowledge Anyway?

Here's something that is uncontroversial (for everyone but the most ardent skeptics, anyway): you know lots of things. For example, you know that you are reading this book right now, you know that the earth is roughly spherical in shape, you know your phone number, you know your name, you know what trees look like, you know that Columbus sailed in 1492, and so on. Here is something else that's uncontroversial (even ardent skeptics agree on this): there are a lot of things you don't know. You don't know who will win each future US presidential election, you don't know whether a fair coin will land heads or tails each time it is flipped, you don't know what color shirts we were wearing while we were writing this book, you don't know on what day the event that led to the extinction of the dinosaurs occurred, and so on. Indeed, there is an enormous number of things that you don't know. Don't feel bad about this—this isn't just true of you; it is true of everyone.

We all know many things, but there are also many things that we don't know. Here, an interesting question arises: What is the difference between knowing and not knowing?

Traditionally, *epistemologists* (philosophers who specialize in issues related to knowledge) have held that knowledge is a true belief that is held on the basis of sufficiently strong evidence.[2] Let us start with the requirement of truth. You cannot know things that are false. This is why you cannot know that the earth is flat. Even if you believe that it is, even if everyone around you believes that it is flat, that doesn't change the shape of the earth. The earth simply isn't flat. So, you cannot know that it is flat. Now, you might wonder: Didn't people in the past know that the earth was flat? The short answer is "no." They didn't know that the earth was flat because it wasn't flat. They only believed that it was. The difficulty here is that it is really easy to run together two different sorts of questions: *metaphysical/ontological questions* (those having to do with how the universe is) and *epistemological questions* (those having to do with what is reasonable for us to believe about how the universe is). We might be tempted to think that people know false things—or knew false things—such as that the earth is flat because we mistake a metaphysical question for an epistemological question. Simply put, it could be that people were being perfectly reasonable when they believed that the earth is flat. All of the evidence they had could have supported thinking that this was true. So, in their situation, it may be that they *should believe* that the earth is flat, and therefore the answer to the epistemological question is that they were rational to believe that the earth is flat. Nevertheless, the fact that they believed that the earth is flat does not entail that they knew (albeit wrongly) that it was flat. The answer to the metaphysical question of what the shape of the earth is depends on the actual shape of the earth, not on what people believe about it. The earth is not flat, and therefore people could not have known that it is flat even though they might have believed that it is. We cannot know things that are not true; we can only have beliefs about them.

This entails that knowledge requires true belief. But is this enough? No: more is required. Imagine that you are about to roll a pair of fair dice—they are not weighted or biased in any way. You roll the dice, and, without looking to see how they land, you believe that you rolled double sixes. Now, imagine that you are in fact correct—both dice actually landed with six dots facing up. Did you know that you rolled double sixes? It seems not. After all, you did not see that this was the case; you merely guessed. You had no good reason to believe that this would be what you rolled. Actually, the odds were heavily against you rolling double sixes: the probability is 1 in 36. However, after looking at the dice, you would know that you rolled double sixes. What's the difference? In the first case, you did not have sufficiently strong evidence to believe that the dice came up double sixes because you did not look and simply guessed—even though the dice did actually land that way! In the second case though, you did have sufficiently strong evidence because you saw the dice with your own eyes. Knowledge therefore not only requires that what you believe is true but also that you have good evidence in support of your belief. Returning to the earth example, we know that the earth is roughly spherical and not flat because we have strong evidence in support of this.[3]

Hence, it seems clear that knowledge requires having strong evidence. But does it also require certainty? This is not a simple question. To answer it, we need to distinguish between two kinds of certainty: epistemic and psychological. *Psychological certainty* concerns how strongly we believe something. We are psychologically certain when we are completely convinced that something is the case, beyond all doubt. Does knowledge require psychological certainty? Can we know things even if we are not completely convinced that we are correct? Indeed, we can. We can have knowledge without psychological certainty; that is, we can know things even if we have some doubt about whether we are correct. For instance, think of someone who is particularly anxious. He always tends to question whether he is correct or whether he has found all the possible information about a topic. Despite his diligent work, he always feels that he is missing something. Would this mean that he does not know anything? It doesn't seem so. Furthermore, being psychologically uncertain might actually make one keen to look for more evidence in support of what one knows. So, psychological uncertainty might actually help build a stronger foundation for one's knowledge.

Importantly, even if we think that knowledge requires psychological certainty, this kind of certainty appears to be independent of the amount or the strength of evidence one has. The more psychologically certain we are about something, the harder it is to reconsider it even when the available evidence against it is overwhelming. For an illustration of the possible disconnect between psychological certainty and the available evidence, think about the beliefs of members of the Flat Earth Society.[4] Despite vast amounts of evidence that the earth is not flat, these people continue to be completely convinced of their view. Their confidence is disconnected from the available evidence; they erroneously believe that they have found sufficiently strong evidence to support their belief that the earth is flat and that "the globe earth theory [is] incoherent with observed phenomena."[5] Hence, psychological certainty isn't really the issue when it comes to how good our evidence has to be in order to have knowledge. One can be psychologically certain about something and still simply be wrong.

What about *epistemic certainty*? Does knowledge require epistemic certainty? Well, before we can answer this question, we need to clarify what epistemic certainty is. Roughly put, you are epistemically certain of something when your evidence is so strong that it makes it *impossible* that you could be wrong. If there is one thing that we should be close to psychologically certain about, it's that humans are fallible. So, it's (almost) always *possible* that we could be wrong about the things we believe. In what follows we consider a number of reasons for thinking that we are not epistemically certain about much, if anything. Of course, this means that *if* knowledge requires epistemic certainty, we are ignorant of almost everything. We come back to this very big *if* at the end of the chapter.

Uncertainty and Perceptual Illusions

We have epistemic certainty when there is absolutely no way that we could be wrong given the evidence that we have. However, our perception of evidence strongly depends on our own senses, which sometimes can deceive us. Our sensory perceptions can lead us to believe something quite confidently only to later discover that we were wrong. These aren't always cases where we lack good evidence. For instance, take a look at Figure 1.1.

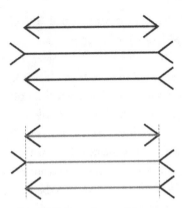

FIGURE 1.1 Müller-Lyer illusion (https://commons.wikimedia.org/wiki/File:Müller-Lyer_illusion.svg).

This figure illustrates the Müller-Lyer illusion. The middle line looks longer than the other two despite the fact that they are the same length (you can measure them with a ruler to verify this). But why does it look longer? It is the direction of the fins that creates the illusion. When they point inward (middle line) they create the impression that the line is longer than when they point outward (top line). Therefore, even though the lines seem to be of unequal lengths, one cannot be epistemically certain that the lines are of unequal length because when one measures them, one will find out that they are not.

Consider another example that shows why evidence from our senses does not make a belief epistemically certain. If you visit Ripley's Believe It or Not museum in London, at some point in the tour there is a rotating tunnel that you have to go through, passing over a bridge. Whereas the tunnel is rotating, the bridge is completely still. Therefore, you are completely still as well, and normally you would be able to cross the bridge walking straight ahead without a problem. Yet the rotating tunnel around you creates the illusion that you are rotating, too, and makes it impossible to walk straight without falling to the sides—even though you are not actually rotating.[6] In this case you have good evidence from your senses that you are rotating, but actually you are not. You might thus be psychologically certain that you are rotating, for instance, if you entered the tunnel without realizing what was going on. But you cannot be epistemically certain about this because you are not really rotating. Once again, sensory evidence fails to make a belief epistemically certain.

It is exactly these sorts of worries about our senses leading us astray that have led philosophers to struggle with external world skepticism (the view that we don't really know anything about the world around us). One of the things that has made external world skepticism so difficult to put to rest is the fact that a skeptic is on to something when he argues that we cannot be certain of the things we take for granted about our everyday lives. We did just see that our senses can be deceptive, after all! A classic presentation of this skeptical challenge was put forward by the French philosopher René Descartes. He employed a method of universal doubt whereby he tried to expose as many uncertain beliefs as possible at the same time. The way that Descartes sought to show that his beliefs were uncertain was to see if he could come up with a *possible* scenario where he had all of the same evidence that he did, but his beliefs were mistaken. The two primary scenarios that he came up with in his famous book, *Meditations on First Philosophy*, were what we might call the "Dreaming Hypothesis" and the "Demon

Hypothesis." The Dreaming Hypothesis is the idea that it is possible that you are dreaming right now. After all, couldn't you have a vivid dream that you are reading this book even though you are actually asleep? The Demon Hypothesis is the idea that it may be that, instead of actually reading this book, you are being tricked by some super-powerful demon into thinking that you are reading it. It is worth pausing here to emphasize that Descartes did not think that either of these hypotheses were likely to be true. He merely accepted that they were *possible*. This is important because if these situations really are *possible* (in the broadest sense of the term), then your evidence for something obvious—such as that you are reading this book—does not make it epistemically certain for you.[7]

Are such skeptical scenarios possible? It seems so. There is no contradiction in thinking that although you have the evidence that you do about something, you are in fact dreaming or being deceived. If these hypotheses are hard to get a handle on, think about popular movies that illustrate similar skeptical scenarios. For instance, in the 1998 film *The Truman Show*,[8] the main character, Truman Burbank, was adopted by a corporation and raised on a film set. All of the people that Truman had met throughout his life were actors playing roles in the show that revolved around him. For a large portion of the movie Truman was completely oblivious to what was going on. This sort of deception does not undercut his evidence for thinking, for instance, that he was reading a book when he was, but it does show that, for a huge portion of his beliefs, he was mistaken despite having the same sort of evidence that we all have for similar beliefs. Whereas he thought that he was living his life like anyone else, he was in fact participating in a show in which he was the only one fooled into thinking that what was happening was real.

Therefore, the evidence for something even seemingly straightforward—such as that you are reading this book right now—does not give you epistemic certainty that you are actually doing so. The same is true for pretty much everything we know, or take ourselves to know, about the world around us. The evidence we have in support of most that we believe does not give us epistemic certainty. We could have that same evidence that we have now even if we were massively deceived because we were in a computer simulation, or dreaming, or being tricked by a demon, or . . . (here you can consider any other crazy possibility). If you are thinking, "this is ridiculous—of course, I know that I'm reading a book!," we agree. You do know that you are reading a book. But still, unlikely as it is, it is possible (perhaps the probability is ridiculously low, but it is still possible) that although it looks like you are reading a book, you are not really doing so. However unlikely it might be, it is possible to be deceived.

Uncertainty and Human Reasoning

It's clear that our perceptions are sometimes mistaken, but what about our reasoning? Can't we be epistemically certain that when we reason to a particular conclusion we are correct? Again, it is important to emphasize that epistemic certainty means that there is *no way* we could be wrong given the evidence that we have. Is it really plausible that humans can reason so well that there is no way that we can be wrong? Well, it doesn't seem so. Our reasoning is fallible, especially when we have to deal with uncertainty.

Let us take a little test. Consider the following description of Linda[9]:

Linda is 31 years old, single, outspoken, and very bright. She majored in philosophy. As a student, she was deeply concerned with issues of discrimination and social justice, and she also participated in anti-nuclear demonstrations.

Now, based on the description of Linda, rank each of the following claims about her from the most probable to least probable:

1. Linda is a teacher in an elementary school.
2. Linda works in a bookstore and takes yoga classes.
3. Linda is active in the feminist movement.
4. Linda is a psychiatric social worker.
5. Linda is a member of the League of Women Voters.
6. Linda is a bank teller.
7. Linda is an insurance salesperson.
8. Linda is a bank teller and is active in the feminist movement.

Do you have your list? Did you rank 8 as more probable than 3 or 6? If so, you committed the *conjunction fallacy*. When two events are independent (such as being a bank teller and being involved in the feminist movement), the probability of them both being true is less than the probability of just one of them being true (for example, if the probability of each event is 0.5, or 50%, then the probability of them both occurring is $0.5 \times 0.5 = 0.25$, or 25%).

So, 3 and 6 are each more probable than 8. However, because of the description of Linda, it can seem more likely that Linda is a feminist bank teller than just a bank teller. About 89% of the people in the original study made this mistake when thinking about Linda's profession.

Let's try one more test. Imagine that you are looking at four cards. Each one has a number on one side and a letter on the other (see Figure 1.2).

The rule is that if a card has a vowel on one side, then it has an even number on the other side. Which cards do you need to turn over to determine whether this rule is true? Keep in mind that you do not want to turn over any more cards than you need to.

The correct answer is card 1 and card 4. The reason is that card 1 has a vowel on it, so if there is not an even number on the other side, the rule is false. Card 4 has an odd number on it, so if there is a vowel on the other side, the rule is again false. Card 2 is irrelevant because the rule doesn't say anything about what happens when there is a consonant on one side of the card. Card 3 is also

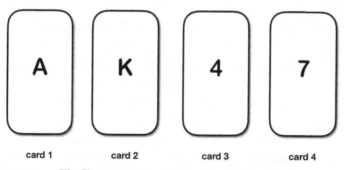

FIGURE 1.2 The Wason selection task (or four-card problem).

irrelevant because, since there is an even number on it, as far as the rule is concerned it does not matter what's on the other side. This test item comes from psychologist Peter Wason's selection task experiment.[10] This sort of experiment has been run a number of times and on average fewer than 5% of people give the correct answer.[11]

Not only do we tend to make errors when reasoning probabilistically (as in the Linda case) and when reasoning about conditional rules (as in the selection task), we are also prone to a large number of cognitive biases. One of the most pervasive is what's known as *overconfidence bias*: we tend to be overconfident of our abilities. As psychologist Thomas Gilovich explained, a "large majority of the general public thinks that they are more intelligent, more fair-minded, less prejudiced, and more skilled behind the wheel of an automobile than the average person." Of course, it can't be that the majority of people are better than average! A further illustration of the pervasiveness of overconfidence bias comes from a study of a million US high school students. This study found that of these 1 million students "70% thought they were above average in leadership ability, and only 2% thought they were below average. In terms of ability to get along with others, *all* students thought they were above average, 60% thought they were in the top 10%, and 25% thought they were in the top 1%." Unfortunately, this bias is not one that afflicts only the young or even those who haven't pursued higher education. A survey of college professors found that "94% thought they were better at their jobs than their average colleague." Our confidence often outstrips our evidence.[12]

Another bias that is particularly relevant to whether we are epistemically certain about the conclusions we draw is what cognitive scientists Hugo Mercier and Dan Sperber call "myside bias." This is a bias that Mercier and Sperber contend is simply a feature of human reasoning. Myside bias refers to the fact that we find it difficult to appreciate evidence that runs counter to our own opinions. As they explain, "Reasoning does not blindly confirm any belief it bears on. Instead, reasoning systematically works to find reasons for our ideas and against those we oppose. It always takes our side."[13] Hence, not only do we often not have all of the evidence, but it seems that in many (perhaps most) cases our cognitive faculties are apt to lead us to interpret the evidence in a way that is colored by our prior convictions. This is why the results of our reasoning do not provide us with epistemic certainty.

But perhaps these are not really errors. Psychologist Gerd Gigerenzer, director of the Harding Center for Risk Literacy at the Max Planck Institute for Human Development in Berlin, has argued that studies like those just described fail to show that we really do fall prey to systematic cognitive errors. According to Gigerenzer, the problem with these studies is that they ignore important pragmatic and context-dependent features that play a key role in human reasoning. The flaw of studies purporting to show that humans are systematically irrational in various ways is that such studies draw conclusions on the basis of whether participants follow particular norms, which may be abstract and incomplete. For instance, regarding the case of Linda being a feminist bank teller, the problem may not lie in how we think but rather in the polysemy of terms such as "probability" that might mislead us. Gigerenzer and his colleagues tested whether the term "frequency" could narrow down the possible spectrum of meanings to those that follow mathematical probability. What they found was that when participants were asked for "frequency" judgments rather than "probability" judgments, they did not commit the conjunction fallacy described earlier.[14] Similar arguments can be made for the Wason selection task.[15] It is important, according to Gigerenzer, not to mistake judgments under uncertainty as cognitive errors. That being said, even those like psychologist Daniel Kahneman who think that we do make systematic cognitive errors do not argue that we should not trust our reasoning abilities. As Kahneman has written, these sorts of errors do not "denigrate human intelligence, any more than the attention to diseases in medical texts denies good health. Most of us are healthy most of the time, and most of our judgments and actions are appropriate most of the time."[16] What matters, as Gigerenzer has argued, is that we can learn to better understand uncertainty so that we are less prone to misjudge the odds of various outcomes.[17]

As a result, the evidence that we have when we reason to a particular conclusion does not make that conclusion epistemically certain for us. Even if we do not in fact make systematic errors in reasoning due to bias or something else, we could be making some other mistake on any particular occasion. Insofar as mistakes are possible, we cannot be epistemically certain. Perhaps then Benjamin Franklin's claim in the epigraph is wrong—maybe even death and taxes are not epistemically certain? People certainly dislike both, but taxes can be avoided in various ways and thus there is no certainty that they will be paid. And, while the fact that this life will end one day is as close to certain as about anything can be, when and how we will die is highly uncertain. Yet even the ultimate end of life is not epistemically certain because there is uncertainty concerning what happens after death. Epistemic certainty is an impossibly high standard to require for knowledge!

What Does This All Mean?

Our sensory perceptions mislead us at times. It is (at least theoretically) *possible* that we are the victims of mass deception. And it is possible to make reasoning errors. Together these facts yield a strong case for thinking that we lack epistemic certainty for many of our everyday beliefs. But what does this mean? Do we simply become skeptics and accept that we know *nothing* about the world around us? No. The uncertainty in our everyday lives simply reveals something important about knowledge: it does not require epistemic certainty. You know that you are reading this book even though your evidence does not rule out every remote *possibility* that you might be wrong. Now, some might insist that knowledge really does require epistemic certainty. That is fine. People can use the word "knowledge" in various ways. However, if knowledge really requires epistemic certainty, then we do not know much of anything. What really matters, regardless of how we use the word "knowledge," is how good our evidence is. As we have seen, our evidence for even the most mundane beliefs in our everyday lives is not enough to make those beliefs epistemically certain. But so what? We make reasonable decisions without being certain all of the time. Some of those decisions are better than others because they are made on the basis of better evidence. Our best bet in any situation is to believe according to what our evidence favors, even if this does not give us epistemic certainty.

For example, if you see that the local weather forecast is predicting a 95% chance of rain, it would be smart to take an umbrella. Are you certain that it will rain? No. Are you even certain that the chance of rain is 95%? No: there is a chance you misread the forecast, or that the meteorologists miscalculated, or that some other mishaps have occurred. Still, it would be a good decision to bring an umbrella with you. Your evidence supports thinking it will rain. Take another example. You have just purchased a lottery ticket for the Mega Millions lottery. The odds are overwhelmingly in favor of you losing (for a January 2018 drawing of this lottery, the odds of winning were 1 in 302.6 million—by way of comparison, the odds of dying because of a lightning strike are about 2,000 times higher than these odds).[18] Despite the odds, you're not epistemically certain that your ticket will lose. After all, there is that 1 chance in 302.6 million that you will win. Nevertheless, it would be unwise for you to quit your job and start planning an extravagant vacation in anticipation of your winning. In both cases, certainty is not required for you to make a good decision: your evidence makes it exceedingly plausible to think that you had better take an umbrella with you and that you will not be winning the lottery. We face uncertainty at every turn in our daily lives. Yet we make good decisions based on the evidence that we have.

Let us now explore the psychology of certainty and uncertainty.

2 | The Psychology of (Un)Certainty

Despite how certainty feels, it is neither a conscious choice nor even a thought process. Certainty and similar states of "knowing what we know" arise out of involuntary brain mechanisms that, like love or anger, function independently of reason.

—*Robert A Burton*[1]

The Illusion of Certainty

You wake up in the morning. Would you like to see your spouse, children, or pet gone without having a clue as to where they went? Would you like to be leaving your house and realize that your car is missing, or wait at the bus stop for 30 minutes for a bus that still has not come? Would you like to arrive at work unsure whether you will be fired? We assume that the answer to all these questions is: No! We all want to know where our spouse, children, pet, or car are; we want to know when the bus will make it to the bus stop; we want to know that we will be able to earn a living in the future. Not only do we want to know these things, we also want to be certain about what will happen. We make plans for the future, from short-term plans about summer vacations to long-term plans about where we want to spend our retirement. We take medical tests because we want to know our health condition and plan for medical intervention if necessary. We like surprises only when they are happy ones: getting an unexpected good job offer or running into a friend we have not seen for a long time. But we hate bad surprises such as being fired because the company decided to downsize or learning about the sudden death of a loved one.

Uncertainty makes us uncomfortable because when we are not sure what to expect it can be difficult to know how to prepare ourselves. People have strived for certainty throughout the centuries through oracles, horoscopes, and other superstitious endeavors. The wish to see into the future and know what will happen gave rise to a number of practices that attempt to fulfill this wish. One example is astrology. Astrology was initially a serious pursuit, and it is no coincidence that notable astronomers such as Tycho Brahe and Galileo Galilei, among others, were involved in astrological activities. Celestial influences were thought to imprint a child with certain properties at the moment of her or his birth, which would in turn have an impact on her or his life. But the aim of astrologers was not to predict anything specific about the future; rather, their aim was to identify possible tendencies or inclinations so that some kind of action could be taken. For this purpose, they constructed horoscopes, which contained the calculations of the positions of all celestial bodies relative to the horizon at exact times.[2] But this is far from today's astrology, which claims to make predictions about one's future. Astrologers now claim to make inferences about one's character and future from charts and horoscopes. Astonishingly, many people seem to believe these claims. For instance, a 2005 Gallup poll in the United States, United Kingdom, and Canada showed that about 25% of participants believed that the positions of the stars and planets affect people's lives.[3]

Another means by which people have tried to remove uncertainty from their lives throughout history is religion. A reasonable argument can be made that religion can provide effective ethical guidance. For instance, one central command in Christianity is to "do to others what you would have them do to you" (commonly referred to as the "Golden Rule").[4] This is ethical advice we can all appreciate, and indeed, if everyone did this, the world would be much better off. Clearly, religion may help one lead an ethical life. It can also help people dealing with pain, and it can provide a source of hope to people in the midst of various struggles. But at times people may take religion to provide epistemic certainty about outcomes that are far from certain—sometimes even unpredictable. Unfortunately, bad things happen to good people, and we often do not know why. Looking to religion or religious texts as a means of being certain about the future is a mistake. In fact, some religions straightforwardly claim that we cannot know what will happen in the future with certainty. For example, in the Judeo-Christian tradition one is told "Do not boast about tomorrow, for you do not know what a day may bring"[5] and "you do not even know what will happen tomorrow."[6]

While we acknowledge that religion can be a positive force in many people's lives and that it may even acknowledge uncertainty, at times religion (like any other belief system) can lead to extreme forms of fundamentalism that provide people with a false sense of certainty that often has dangerous consequences. Many religious people are psychologically certain of the correctness of their faith. Unfortunately, there are those who allow their psychological certainty of their own correctness to lead them to condemn, or even harm, those who disagree with them. It is a sad fact that instances of religious fundamentalists doing bad things are not hard to find. Take the actions of the Westboro Baptist Church or fundamentalist Islamic groups such as Al-Qaeda, for just two straightforward examples. More generally, throughout history, there have been a number of religious conflicts resulting in the deaths of numerous people. In many cases all involved were certain that they were right and that those they were fighting were wrong. But what leads people to do horrible things in the name of religion (or national pride, or whatever)?

No doubt there are a variety of factors, but perceived certainty seems to play a big role. As anthropologist James Houk has put it, one thing that leads people to harm others in the name of religion is "the illusion of certainty that is borne of religiosity of the worst sort, namely religious fundamentalism. To the fundamentalist, everything is black and white, the truth is absolutely known (to them) and absolutely nonnegotiable."[7] The same is true of other forms of fundamentalism. In its various forms (religious, political, and so on), fundamentalism is problematic because it rests on a misunderstanding of the world and our limitations as humans. It is impossible to be epistemically certain about much of anything, regardless of our felt need to be psychologically certain. Arguably, the only certainty in human life is death; but even when this will occur and what happens after is highly uncertain. We cannot be certain about what the future will bring, even though we can do a lot to achieve the outcomes we wish by using the evidence we have to make reasonable predictions. But why do we feel the need to be certain?

The Psychological Need for Certainty

To understand the psychological need for certainty we need to understand how the mind works. Neurologist Robert Burton has explained that we can find ourselves in some strange situations concerning what we actually know and what we think we know. On the one hand, it is possible to really know something without having the feeling of knowing, the psychological certainty that we know it. Burton has described a situation that clearly exhibits this state; because of a stroke, patients with *blindsight* have a destroyed occipital cortex—the part of the brain that receives visual inputs—but they can accurately localize a flashing light even though they are not conscious of seeing it. This happens because neural fibers other than those involved in conscious vision experiences send a stimulus to areas of the brain that do not produce a visual image but are involved in automatic, reflexive reactions. Because of this stimulus, the brain of the person perceives the light without the projection of a visual image.[8] In other words, these people can know that there is light somewhere without being aware that they actually know this. A more mundane example of this phenomenon occurs when someone knows something but has simply never reflected on the question of whether they know it. For example, when you are walking down the street, you know a whole host of things: cars are passing by, the sidewalk is uneven over there, a bird is flying above you, etc. But for a vast number of these things that you know you never reflect on whether you know them—you simply process the information and act accordingly. Hence, for a sizeable portion (perhaps most) of the things you come to know while walking down the street you do not have a feeling of knowing them.

The opposite situation is quite common as well: having the feeling of knowing something without actually knowing it. People can be committed to some beliefs to the extent that they are so psychologically certain about them that they do not abandon them even in the face of contradictory evidence—this is a state of cognitive dissonance. In such cases people may believe that they correctly remember the details of an event, whereas they are actually wrong about these. They may feel that a pseudo-medical treatment cured their illness, ignoring the placebo effect; or they may stick to unfounded beliefs such as the idea that God created the world in its present form while ignoring the overwhelming evidence for geological changes and the evolution of life that have taken place on Earth.[9] In other words, they can claim to have knowledge that they do not really have even when facing strong evidence that they are mistaken. People can be fully convinced that their beliefs are correct even though they are completely wrong.

But how does this work? Do the people with blindsight have special powers; are proponents of creationism irrational or stupid? It seems that the feeling of knowing is something we do not control. Imagine the following situation: you are alone in a dark room in a place where you have never been before. Suddenly you hear a noise right behind you. Before you turn around to see if anyone is there you feel your body trembling. You feel fear. But you realize it after you have felt it; it is not a conscious feeling. You did not weigh the evidence and think to yourself "that noise must have been caused by someone else in the room, therefore I should be concerned because perhaps I am in danger." You simply had an unconscious reaction to your situation. Anger is similar in this respect; it can come on us without conscious reflection. Imagine, you are crossing the street and you hear a car approaching at high speed, completely disregarding the speed limit in your neighborhood and the fact that you have the right of way since you are in the crosswalk. You stop, and you see the driver passing in front of you. He is oblivious to what is going on because he is busy talking on his phone. You feel your blood pressure rise, and your head feels like it will explode. You are angry. But again, you only realize this after it happens. Both fear and anger are primary, involuntary feelings, and it seems that this is also the case for the feeling of knowing. Such a feeling has been linked to a discrete anatomic localization in the brain, and it can be easily reproduced without conscious cognitive input. It just happens, unconsciously.[10]

But if we cannot control what we feel we know then can we really know anything? Yes, we can, and we do. As Burton has put it "in pointing out the biological limits of reason, including scientific thought, I'm not making the case that all ideas are equal or that scientific method is mere illusion. . . . The purpose is not to destroy the foundations of science, but only to point out the inherent limitations of the questions that science asks and the answers it provides."[11] Philosophers have long denied what is called the "KK" principle, which says that in order to know something you have to know that you know it. There are very good reasons for denying this sort of principle in general. Plausibly, animals and young children know a number of things, at least about what is going on around them when they are conscious. But do they know that they know? It seems not. Animals and young children do not even have the concept of "knowledge," so they cannot form beliefs about what they know. As a result, they cannot know that they know (or do not know) something. Consequently, the mere fact that we cannot control our feelings of knowing does not imply anything about whether we actually know things.

How do our brains deal with uncertainty? Let us consider a study aimed at investigating neural responses to visual cues indicating the occurrence of painful stimuli with different levels of uncertainty. Twenty-five participants (13 women and 12 men, 19–36 years old) underwent two procedures at separate times about 50 days apart: (1) electroencephalography (EEG), which measures neural processes with high accuracy, and (2) functional magnetic resonance imaging (fMRI), which allows the assessment of neural activation with high spatial resolution. The aim was to explore the differences in neural processes between psychologically certain and uncertain conditions during the anticipation of pain as well as during the delivery of pain. A visual cue indicated the occurrence of a painful or nonpainful stimulus, which was delivered during the subsequent 15 seconds. There were four stimulation conditions: certain pain (a painful stimulus had been announced with a probability of 100%), certain no pain (a nonpainful stimulus had been announced with a probability of 100%), uncertain pain (a painful stimulus had been announced with a probability of 50%), and uncertain no pain (a nonpainful stimulus had been announced with a probability of 50%). It was found that during the first 2 seconds of pain anticipation, the uncertain cue was associated with stronger affective anticipation processes and captured more visual attention than the other cues. It was also found that during the last 2 seconds before stimulus delivery, uncertain anticipation was associated with attentional control processes. Finally, it was found that unexpected painful stimuli produced the strongest activation in the affective pain processing network. Overall, uncertainty seemed to activate various areas of the brain, which might have an adaptive value for living in uncertain environments.[12] This is a small study of course, but the results are interesting. A plausible answer to the question of why we strive for certainty, suggested by this study, is that uncertainty is stressful.

Another study aimed at evaluating the contribution of different forms of uncertainty to physiological stress responses also suggests this answer. Forty-five participants (25 women and 20 men, 19–35 years old) completed a task during which they were presented with one of two rocks and were asked to predict whether or not there was a snake underneath it. There was a probabilistic association between rocks and the presence or absence of a snake, which was not explicitly mentioned to participants. During the task the probabilities varied from heavily biased (90/10) to unbiased (50/50) toward one outcome. In other words, the probabilities varied from less to more uncertain, thus allowing researchers to examine the effect of unpredictability on stress responses. Each time a snake was presented, participants received a painful electric shock to the hand. The researchers measured cortisol levels in saliva, as well pupil diameter and skin conductance in order to confirm participants' stress responses. The analysis of the results showed a strong relationship between subjective estimates (participants were asked at regular intervals to rate how stressed they felt) of uncertainty and stress. The researchers also found that both pupil diameter and skin conductance were increased by uncertainty.[13] This is another small study but the conclusion seems plausible: uncertainty is stressful; we dislike feeling uncertain. The less predictable an outcome is, the more stressed we become.

Overall, it seems that uncertainty unconsciously activates our brains and causes us to feel stress. Do you enjoy feeling alerted and stressful more than feeling relaxed and calm? Most likely not. This is why we tend to strive for certainty in all facets of our lives. As the authors of an older study, which provided evidence that our brain has a neural system for evaluating general uncertainty, nicely put it: "Under uncertainty, the brain is alerted to the fact that information is missing, that choices based on the information available therefore carry more unknown (and potentially dangerous) consequences, and that cognitive and behavioral resources must be mobilized in order to seek out additional information from the environment."[14] We have all felt it, haven't we? This is discomforting, and so we prefer to be certain. But absolute certainty is something we cannot achieve, and this is why we need to accept, and learn to deal with, uncertainty.

Living with Uncertainty

If epistemic certainty is an illusion but we have a psychological need for it, what can we do? The answer, in our view, is simple and straightforward: we need to accept uncertainty and learn to live with it. But this can only be achieved on a large scale if people start learning how to do this from a young age. Learning how to deal with uncertainty should become a central goal of our education because we face uncertainties in all aspects of our lives: family, friendships, marriage, career, health, and more. But as psychologist Gerd Gigerenzer has correctly remarked:

> The problem is that our educational system has an amazing blind spot concerning risk literacy. We teach our children the mathematics of certainty—geometry and trigonometry— but not the mathematics of uncertainty, statistical thinking. And we teach our children biology but not the psychology that shapes their fears and desires.[15]

It's as simple as that. Education often does not focus on what people really need to learn but instead on what some normative, if not dogmatic, assumptions say about what people ought to learn. In his own work, Gigerenzer has shown the problems with our illiteracy concerning uncertainty and how to deal with it.

A similar point was made by neurologist Robert Burton:

> I cannot help wondering if an educational system that promotes black or white and yes or no answers might be affecting how reward systems develop in our youth. If the fundamental thrust of education is "being correct" rather than acquiring a thoughtful awareness of ambiguities, inconsistencies, and underlying paradoxes, it is easy to see how the brain reward systems might be molded to prefer certainty over open-mindedness. To the extent that doubt is less emphasized, there will be far more risk in asking tough questions. Conversely, we, like rats rewarded for pressing the bar, will stick with the tried-and-true responses.[16]

Here is the problem in a nutshell: in our schools, at least in the Western world, we learn that things are black or white and that we can always identify the correct answer to any question. But the real world is very different: things are not black or white, but may rather be one of the numerous shades of gray between them. A central goal of education should therefore be the appreciation of how our world really is. In some cases, we know the likelihood of various outcomes, but in other cases we simply do not know them—yet or at all. This does not mean that we cannot know anything or that we need to question everything. Rather it means that we need to liberate ourselves from the inherent need to be certain about what will happen and do our best to understand and prepare ourselves for the likeliest outcomes. This is all we can do.

Here is just one example of how even official reports may be misleading because of the public's illiteracy when it comes to probabilities. The probability of a woman developing breast cancer is generally not very high, but for women carrying versions of particular genes, *BRCA1* and *BRCA2* ("BRCA" simply means breast cancer) the probability is relatively high. According to the 2016 report of the American Cancer Society: "Compared to women in the general population who have a 7% risk of developing breast cancer by 70 years of age, the average risk for *BRCA1* and *BRCA2* mutation carriers is estimated to be between 57%– 65% and 45%– 55%, respectively."[17] This simply means that whereas by 70 years of age 7 out of 100 women of the general population will have developed breast cancer, this will be the case for 57–65 out of 100 women with a particular version of *BRCA1* and for 45–55 out of 100 women with a particular version of *BRCA2*. Put simply, whereas fewer than 1 in 10 women from the general population will develop cancer by the age of 70, for women carrying a particular *BRCA1* or *BRCA2* mutation it will be approximately 1 out of every 2 of them. The upshot of this information might seem straightforward, and it has resulted in headlines like the following: "Women like Angelina Jolie who carry the BRCA1 gene are less likely to die from breast cancer if they have their OVARIES removed"[18]; "Study: Women with BRCA1 mutations should remove ovaries by 35"[19]; "Moms with BRCA breast cancer gene mutations face tough decisions."[20]

However, if one reads the American Cancer Society report, earlier in the same paragraph we quoted it is stated: "Inherited mutations (genetic alterations) in BRCA1 and BRCA2, the most well-studied breast cancer susceptibility genes, account for 5%–10% of all female breast cancers." This means that out of the 100 women who will develop breast cancer, only 5–10 will be *BRCA1* and *BRCA2* mutation carriers. If you do the math, this means that women with mutated *BRCA* alleles who will develop cancer by the age of 70 are 5–10% of 7% of the general population (i.e., less than 1% of the general population). What is the conclusion? Women carrying *BRCA1* and *BRCA2* mutations should be concerned, but they are only a small portion of the women developing breast cancer and only a very small portion of the general population. As a result, women should not get a genetic test for the *BRCA* alleles before having other medical tests related to the diagnosis or prognosis of breast cancer unless perhaps there is a history of breast cancer in the family. Failure to appreciate what the probabilities actually tell us can lead to unnecessary stress and expense for women and their families.

What is the take-home message? We cannot really be certain about many aspects of our lives, as much as we would have liked it to be the case. Some of us have a higher probability of developing one disease or another during our lifetime. What we can do is eat healthy, get a sufficient amount of exercise, and take regular medical tests to see if our body functions properly and do something when it does not. More broadly, we cannot be certain whether our marriage will work out, how our children will turn out, whether we will find the job of our dreams or lose it once we have it, how many copies this book is going to sell, whether critics will like it or not, and so on. All we can do is try to make the best decisions we can based on the available evidence and see what life brings without worrying about the rest.

3 | Uncertainty in Science
ISN'T IT A PROBLEM?

It's 46° (really cold) and snowing in New York on Memorial Day—tell the so-called "scientists" that we want global warming right now!

—*Donald J. Trump*[1]

Ought Scientists Be Certain?

Donald Trump is notorious for his preference for turning to Twitter to express his views on various topics, a preference that does not seem to have declined since he became the 45th President of the United States in 2017. Between 2011 and 2015, he expressed skepticism about climate change and global warming on his Twitter account no less than 115 times.[2] In some of those tweets, such as the one in the epigraph of this chapter, he also expressed his distrust for scientists working in this domain. The point that he essentially made was that scientists are talking about global warming when people have to deal with low temperatures (46°F, the equivalent of about 8°C) at a time of the year that is typically quite warm. Therefore, scientists either are getting things wrong, which they are not supposed to do (hence "the so-called 'scientists'" comment), or—worse—they are distorting the truth by manipulating data, as he has claimed elsewhere.[3] The implicit message here is that scientists should have known better and should have been certain about what they are talking about before they made any pronouncements to the public. What they claim about global warming does not align with the everyday perception of low temperatures. Where is the global warming if Donald Trump and all of us are freezing out in the street when it is almost summer time? Aren't scientists certain of the claims they make? If not, then why are they making those claims?

There are two important issues here. The first one is that, as Trump's comment shows, people may be ignorant of what climate change and global warming actually are. Simply put, global warming is not the phenomenon of having high temperatures throughout the whole year and therefore not having low temperatures at all. Rather, global warming refers to the rapid increase in the average temperature of the atmosphere in recent times. The 2014 report on climate change by the Intergovernmental Panel on Climate Change (IPCC), the international body for assessing the science related to climate change that was set up in 1988 by the World Meteorological Organization (WMO) and United Nations Environment Programme (UNEP), has suggested that the human impact on the climate is clear. Recent anthropogenic emissions of greenhouse gases—such as carbon dioxide (CO_2) and methane (CH_4)—are higher now than at any point in history. The warming of the climate is unequivocal, and the changes that this has brought about are significant. In particular, the globally averaged combined land and ocean surface temperature data show a warming of 0.85°C (about 1.5°F) between 1880 to 2012, with the global mean sea level rising by 19 cm over the period from 1901 to 2010. At first sight, this might not seem like a lot. However, as the report indicates, the period between 1983 and 2012 seems to have been the warmest 30-year period of the last 1,400 years in the Northern Hemisphere. And the rate of sea level rise since the mid-nineteenth century has been greater than the average rate of rise during the previous 2,000 years![4] Simply put, the atmosphere is getting warmer, and the sea level is rising at an alarmingly fast pace (more about this topic in Chapter 6).

The second important issue here relates to the certainty that scientists have, or ought to have, in order for the public to trust their claims and recommendations. On April 2017, Bret Stephens wrote his first column for the *New York Times*, "Climate of Complete Certainty."[5] There, Stephens commented on the findings of an October 2016 Pew Research Center report on the politics of climate change.[6] He wrote:

> Just 36 percent of Americans care "a great deal" about the subject. Despite 30 years of efforts by scientists, politicians, and activists to raise the alarm, nearly two-thirds of Americans are either indifferent to or only somewhat bothered by the prospect of planetary calamity.
>
> Why? The science is settled. The threat is clear. Isn't this one instance, at least, where 100 percent of the truth resides on one side of the argument?
>
> Well, not entirely.

According to Stephens about 64% of people in the United States do not care much about climate change because scientists have failed to convince them that this is a really important issue. And the reason for this, Stephens implied, is that scientists are not 100% certain about what is going on, even though they claim to be. He went on to suggest that whereas the "modest" warming of the earth since 1800 mentioned in the IPCC report and the human influence on that warming are indisputable, "much else that passes as accepted fact is really a matter of probabilities." With that, he noted that he did not intend to deny climate change or its possible consequences. Rather, his intention was to question the rather "scientistic"[7] attitude of climate scientists by noting that "Claiming total certainty about the science traduces the spirit of science and creates openings for doubt whenever a climate claim proves wrong." In other words, climate scientists should not claim to possess a certainty they do not actually have because this might make them seem less trustworthy when it becomes clear that there is uncertainty about the details of climate science.

Climate scientists quickly responded to Stephens with an open letter, signed by 137 scientists as of the time of our writing, expressing their concern about the "inaccurate and misleading statements about the science of climate change" made by Stephens. They severely (and correctly) criticized several of his points, such as the claim that the warming of the Earth since 1800 has been "modest." They explained that this characterization is inaccurate and misleading because the warming that has occurred is both significant and rapid, more than *100 times* as fast as the cooling that took place over the previous 5,000 years, and because it can have devastating consequences: "Much as a fever of only several degrees can be deadly, it only requires a few degrees of warming to transition the planet out of ice ages or into hot house conditions." Most importantly, the scientists noted that Stephens both mischaracterized the uncertainties of climate change and misrepresented how science reports uncertainties: "Contrary to the writer's false accusation that scientists claim total certainty regarding the rate of warming, IPCC reports present a range of estimates for global warming—centering around 1°C (about 1.8°F) of warming since pre-industrial times."[8] In other words, climate scientists never claimed to be 100% certain. And no serious scientist would ever make such a claim.

Misunderstanding Uncertainty

Uncertainty in science is not taboo, of course. It is included in public representations of science, for instance in articles about climate change.[9] Additionally, communicating scientific uncertainty to the public may not have detrimental effects on the public's trust in scientists and their work.[10] However, it seems that how people understand the accuracy of a particular scientific field guides their perception of it and how they perceive and evaluate its uncertainty.[11] Thus, undesirable results are often discredited and important policies rejected in the name of uncertainty. Therefore, any attempt at improving the public understanding of science should include a focus on explaining the roots and the possible impact of uncertainty in science. An important distinction that must be kept in mind is that between psychological certainty and epistemic certainty, already discussed in Chapter 1. Psychological certainty concerns our psychological commitment to what we believe, whereas epistemic certainty concerns the strength of our evidence when it comes to a particular belief. Whereas knowledge requires truth, it does not require certainty of either of these two kinds—in fact, epistemic certainty is an unattainable goal for the vast majority of cases. This is especially important to realize because people may favor a science that provides absolutely certain, "black or white" answers when, in fact, the answers usually lie among the 50 (or 500, or 5,000) shades of gray. For instance, it has been found that when TV shows represent science as certain people tend to exhibit increased interest in it, whereas when TV shows represent science as uncertain there may not be such an effect.[12] Science can make testable predictions and confirm or reject hypotheses, but in no way does it provide definitive, absolutely certain answers to all (or perhaps any) of its questions. If it did, there would be no point for further scientific inquiry. But, even though uncertainty motivates further research, it may nevertheless be disparaged and misunderstood by the public. Why is this happening? One reason is that we have an inherent need to feel certain, as we showed in the previous chapter. We therefore may tend to look for certainty around us and feel uncomfortable with its absence.

Perhaps as a result, uncertainty in science has at times been seen as a reason for concern and has often been misused in public controversies about socio-scientific issues, as shown in the case of climate science (the past offers plenty of other examples, such as tobacco smoking and cancer, greenhouse gases and global warming, ozone depletion, and more[13]). Because of this misuse, there has been public discussion about whether there is a consensus within the scientific community concerning the causes of climate change and how large this consensus actually is. Perceived scientific consensus seems to have an impact on the acceptance of scientific conclusions by the public. For instance, a recent study found that highlighting the consensus within the scientific community resulted in increased acceptance of science. Participants were found to be more willing to attribute long-term climatic trends to human activities when they had been informed about the scientific consensus on the topic, and they were more likely to accept as true the statement that human CO_2 emissions cause climate change.[14] A consensus within the scientific community might be taken to imply certainty; in contrast, when there is no consensus among scientists, it could seem as if there are different reasonable views on the topic perhaps because there is more uncertainty about the evidence and the possible interpretations. Unfortunately, the public understanding of important socio-scientific issues has been, and can still be, manipulated through the presentation of lack of consensus and misrepresentation of the nature of uncertainty in science.

Of course, climate change is not the only hotly debated topic where misperceptions about uncertainty lead to misunderstanding the science. Parents may hesitate to vaccinate their children for various reasons, including the compulsory nature of vaccination, their unfamiliarity with vaccine-preventable diseases, and the lack of trust in pharmaceutical companies and public health agencies.[15] Uncertainty is an important underlying factor in this case, too; the uncertainties about the potential side effects of vaccines and more broadly about whether vaccination does more harm than good may impact parents' decision as to whether or not to vaccinate their children (see Chapter 7). Last, but not least, there is uncertainty about human evolution. Whereas scientists have a rich picture of how we have evolved to our present state, this picture is quite fragmented. This poses no questions about the fact of human evolution, but there exists uncertainty about its details— how exactly human evolution took place (see Chapter 8).[16] In all these cases, anti-science arguments by climate change deniers, anti-vaxxers, and anti-evolutionists have questioned the conclusions of scientific research in the name of uncertainty.

At the same time, people tend to perceive particular scientific findings as certain, completely overlooking important uncertainties inherent in the respective procedures, as in the case of genetic tests. Important uncertainties characterize these tests, and these uncertainties are related to whether the test can accurately detect the presence or absence of a genetic variant (analytical validity) and to how well a genetic variant is related to the presence, absence, or risk of a specific disease (clinical validity).[17] These uncertainties are not always clearly explained, and, as a result, it is not always clear to lay people what exactly a genetic test can and cannot reveal and what expectations one should reasonably have from test results. An increased probability to develop a genetic disease does not entail that one will indeed develop the disease, and any prediction for developing genetic diseases is valid only under very specific conditions (see Chapter 9). For similar reasons, forensic tests are not infallible and cannot unquestionably help decide whether a suspect is guilty or innocent. Many factors can mislead the analysts, and different kinds of uncertainty are inherent in the respective procedures, from the origin of the analyzed sample to the conclusions of this sort of analysis (see Chapter 10).

It is interesting that, in the name of uncertainty, people question whether humans have significantly contributed to climate change, that children should be vaccinated, or that humans have evolved by natural processes, but at the same time people overlook the fact that particular genetic tests can be useful only in particular cases and that forensic tests can be misleading. In these and other cases, the problem is not that scientists are uncertain about the answers to particular questions: they are, and this is normal. Rather, the problem is that the uncertainty concerning very specific aspects of an issue is often mistakenly taken to indicate that science has not provided us with sufficient evidence to make it rational to accept the conclusions of scientists. Uncertainty does not mean that we cannot know anything, that all views are equally valid, that all scientific evidence should be questioned, or that important socio-scientific decisions cannot be made. It only means that scientific inquiry is a continuous process, and we have to make the best decisions we can with the available, sometimes limited, though often quite strong, evidence.

Understanding Uncertainty

A first step to making the best decisions we can with the available evidence is to clarify what uncertainty is. Overall, we can be quite sure about some phenomena, even though there is uncertainty in the details. Consider a simple example. If someone throws a big stone at a door with glass panes, it is almost certain that some of the panes will break. Of course, there is always the possibility that all panes will remain intact for whatever reason (e.g., because of their material or because the throw was poorly aimed). But generally speaking it is very close to (but never completely) certain that some of the panes will break. Now, what is particularly uncertain in advance of the throw is which of the several glass panes will break. This is something one can only know after the stone is thrown because one cannot know in advance exactly which pane the stone will hit. Of course, we could create a model that would take into account the initial speed of the stone, the angle at which it will be thrown, and, thus knowing the distance from the door, calculate the velocity with which the stone is going to reach the door panes. We could thus *estimate* that some glass panes might have a higher probability of breaking than others. But, of course, there remains uncertainty as to what will actually happen every time a stone is thrown at the door. Scientists are very close to certain that climate change and human evolution are real phenomena, as well as that vaccinations and genetic and forensic tests can be useful in the same sense that one can be nearly certain that throwing a stone at a door with glass panes will result in some of the panes being broken. However, at the same time there is uncertainty about several details of climate change, human evolution, vaccination, and genetic and forensic testing, similar to the uncertainty concerning which particular glass panes will break.

This uncertainty can be expressed in terms of *probabilities*. We can say that throwing a stone at a door with glass panes has a high probability of breaking one or more of them. How do we know that? We know from experience that heavy objects, like stones, tend to break glass when they strike it. We can also test this experimentally. If we throw the same stone at 100 identical doors with glass panes, we will likely see that the panes in a very high portion of these doors, say 99, will break. On this basis, we could say that the probability that the panes of this particular type of door will break when this particular stone is thrown at them is 99%. But if we look at which panes actually broke, things might become more complicated. We might see that—assuming each door had 4 large glass panes—the top left pane was broken in 15 of the doors;

the top right pane was broken in 20 of the doors; the bottom left pane was broken in 35 of the doors; the bottom right pane was broken in 29 of the doors; and there was one door with no broken panes. On the basis of these results, we could say that the probabilities for the particular stone to break the panes of this particular kind of door are 15% for the top left pane, 20% for the top right pane, 35% for the bottom left pane, and 29% for the bottom right pane. Given these data, we might even develop a model that offers predictions for different types of stones or doors. But still, any prediction would be probabilistic—the odds of any particular outcome would be less than 100%. In any case, the point made here is that even if we cannot be completely certain of any of the outcomes, we can be more or less uncertain about some outcomes compared to others.

A related and important distinction has been proposed by Gerd Gigerenzer, who has distinguished between the probabilities that we can know in advance and those that we cannot know in advance. The first kind of probabilities he has called "known risks" or simply "risks," whereas he has referred to the second kind as "unknown risks" or "uncertainties."[18] This distinction is very important because it makes a big difference both psychologically and epistemically whether the probabilities of an event's occurring or not are known in advance. In this book we refer to this distinction with a slightly different terminology, talking about *known* and *unknown uncertainties*, respectively. We believe that the terminological contrast between *known* and *unknown* uncertainties is clearer than that between *risk* and *uncertainty*, but, of course, the underlying concepts are essentially the same. Let us see in more detail what these are about.

Known uncertainties arise in situations where we can know in advance how probable particular outcomes are (hence the designation *known*), but we cannot know in advance which one among several possible outcomes will actually occur (hence the *uncertainty*). For instance, when you flip a fair coin, you know in advance that the probability of getting heads or tails is 50% (assuming that the coin is flipped fairly). Nevertheless, even if you know this, you cannot tell when you flip the coin whether you will get tails or heads. Similarly, if you buy 1 lottery ticket in a fair lottery with a guaranteed winner, and you know that 10,000 tickets were sold, you can know that the probability of your winning is 1 in 10,000 or 0.01%. If you decide to buy 100 tickets, then you know that the probability of winning is 100 in 10,000 or 1%. Thus, even if you cannot know in advance whether the winning ticket is among those you have purchased, you can know how likely this is. In contrast, *unknown uncertainties* arise in situations where we cannot know in advance how probable given outcomes are (hence the designation *unknown*), and we cannot know in advance which one among several possible outcomes will actually occur (hence the *uncertainty*). This would be the case if you bought a ticket for a lottery for which you did not know how many tickets were made available. If the tickets sold were 10, then by buying 1 ticket your chances of winning would be 10%. If the tickets sold were 1,000, then your chances would be 0.1%. But insofar as you do not know how many tickets were sold, you cannot know whether it is relatively likely or relatively unlikely that you will win the lottery.

This raises a very important related point: dealing with uncertainty depends on our understanding of probabilities. However, there is evidence that people intuitively misunderstand probability.

Gigerenzer conducted a study asking 750 people in Amsterdam, Athens, Berlin, Milan, and New York what the statement that there is a "30% chance of rain tomorrow" means. In some cases, people were asked to give their own response; in other cases they were asked to choose from among the following three options: (1) it will rain tomorrow in 30% of the region, (2) it will rain tomorrow for 30% of the time, (3) it will rain on 30% of the days like tomorrow. The researchers found that only in New York did the majority of participants select option 3, which is the standard meteorological interpretation. In each of the other cities, this option was judged as the least appropriate. The preferred interpretation in the European cities was that it will rain tomorrow "30% of the time," which was followed by "in 30% of the area."[19] This simple case clearly illustrates some of the problems with the interpretation of probabilities. A main issue, Gigerenzer noted, is that people do not understand probabilities because the reference class (percentage of what?) is not always clearly communicated to them.[20] As we will see in other chapters of this book, misunderstanding probabilities is a major reason that people fail to understand the uncertainty inherent in science. We are not really educated to deal with uncertainty, as Gigerenzer has noted. It is, therefore, necessary that education makes understanding uncertainty and probabilities a priority, as well as that scientists learn how to communicate their probabilistic findings to lay people in a way that the latter can make sense of them.

With these distinctions in mind, let us now take a look at what people generally think about science and its trustworthiness.

4 | In Science We Trust—Or Don't We?

I can live with doubt, and uncertainty, and not knowing.
I think it's much more interesting to live not knowing than
to have answers which might be wrong. I have approximate
answers and possible beliefs and different degrees of certainty
about different things. But I'm not absolutely sure of anything, and there are many things I don't know anything about.
—*Richard Feynman, BBC interview*[1]

Expertise in Science

Richard Feynman might seem like the exemplar of a twentieth-
century scientist. He was involved in the Manhattan Project that
resulted in the development of the atomic bomb in the 1940s. In
the 1950s and 1960s, he was nominated 48 times for the Nobel
Prize in physics,[2] which he was eventually awarded in 1965, jointly
with Sin-Itiro Tomonaga and Julian Schwinger "for their fundamental work in quantum electrodynamics, with deep-ploughing
consequences for the physics of elementary particles."[3] He was also
a very famous popularizer of science, writing several widely read
books.[4] One example of Feynman's scientific prowess occurred in
1986, when the space shuttle Challenger exploded a few moments
after its takeoff, resulting in the deaths of its crew. Feynman joined
the commission that was formed to investigate the accident. He

conducted diligent research on the topic, and he eventually figured out that the rubber in the O-rings used to seal the solid rocket booster joints failed to expand when the temperature was at or below 32°F (0°C), which was the temperature when the accident happened. In a spectacular moment, Feynman demonstrated this by putting such a ring into ice water. He then explained that because the O-rings cannot expand at 32°F, there were gaps in the joints and so the gas got out, which led to the explosion of the booster and eventually of the whole shuttle. With a simple experiment, Feynman expertly showed the most likely cause of that accident.[5]

But then how could this brilliant scientist state, as the epigraph in this chapter indicates, that he was not absolutely sure of anything and that there are many things he did not know? Shouldn't one expect exactly the opposite from him? The answer is a definitive "No!" In fact, it is exactly because Feynman was both brilliant and knowledgeable that he was able to realize that there were topics beyond his area of expertise that he knew nothing about and that he was also aware that, even for those topics he knew a lot about, absolute certainty was not possible to achieve. This raises two important points: first, Feynman knew a lot about particular topics but not about all topics; second, and most important for the purposes of this book, even Feynman did not know everything about the particular topics he knew a lot about. This entails that Feynman—and every scientist—can be an expert in particular areas and know a lot without knowing everything.

But how does one qualify as a scientific expert? We consider scientific experts those people who have mastered science-related knowledge and skills, who practice these as their main occupation, and who have valid science-related credentials, confirmed experience, and affirmation by their peers. Even though members of the public may occasionally be involved in science-related activities, they do not do this as their main occupation and also lack the knowledge, skills, credentials, experience, or peer affirmation that scientists have. This, of course, is not to deny that there are scientists who are not very good at what they do; this is true of some, but even they are less likely to reach wrong conclusions when it comes to their fields than nonexperts.[6] Most importantly, what matters for us in this book are not the views of any individual expert, but the consensus views and the conclusions that the majority of experts in scientific communities have reached. Throughout the book we consider the views that the majority of experts in a particular domain hold to be the most accurate view currently available. If we have communities of people whose main occupation is to engage in scientific research and who have the skills, knowledge, credentials, experience, and peer affirmation for doing this, then these are the most competent people to know what can be known about the topics that it is their job to study.

The preceding definition of "expert" implies that the public ought to trust experts. But what exactly is "the public"? It is far from simple to define what "the public" is. In this book we use this noun vaguely to refer to all ordinary people who are not experts in science. We thus contrast them with scientists, who are the experts, even though scientists are, of course, citizens as well. Since experts are the most likely to have the relevant knowledge, they are the ones the public should listen to when it comes to the topics of their expertise. Of course, even though scientists are experts in their specific domains, they do not know everything about them.

Furthermore, experts in one domain—say, climate scientists—are not experts in other domains; a climate scientist should be considered a member of the public when it comes to vaccination or any other topic beyond his or her area of expertise. Usually, experts are well aware of what they do not know and what they might never know. Being an expert means being aware of both the current knowledge on a topic and its limitations. We thus start from the premise that climate scientists are those who know better than anyone else about climate change; pediatricians, pathologists, and other medical doctors are those who know better than anyone else about vaccines; evolutionary biologists are those who know better than anyone else about evolution; geneticists are those who know better than anyone else about genetic testing; and forensic experts are those who know better than anyone else about forensic testing. It is the consensus views of these people that the public needs to listen to, keeping in mind that the community of experts may be debating the details among themselves, as we show in subsequent chapters. Trusting the opinions of these experts results from the same rationale that we follow when we call a mechanic when our car does not start instead of trusting our own intuitions and when we let the pilot fly the plane that takes us wherever we want to go rather than demanding to fly it ourselves. Pilots know more about planes and how to fly them, and mechanics know more about cars and how to fix them than the average person. They are experts in their domains. In the same sense, scientists know more about natural phenomena of various kinds than the public. They are the experts when it comes to science.

Science is by its very nature a self-correcting enterprise that constantly advances. This entails that some current scientific claims will be modified, or even discarded, at some point in the future, but it also entails that scientific knowledge is continuously being established on an increasingly solid evidential basis. Indeed, there exist several scientific theories of the past that have been rejected or that have evolved into newer theories. This provides evidence for thinking that the same could also happen to our current scientific theories. But this does not devalue those replaced or rejected theories because, for as long as they were accepted, they were the best available, providing significant understanding of their respective phenomena. The conclusions of good-quality science provide the best rational understanding of natural phenomena available at a given time, and we had better accept it (although science is not of good quality all the time, an issue discussed in the next chapter). The history of science is not a wasteland of false theories. As philosopher Stathis Psillos put it: "The replacement of theories by others does not cancel out the fact that the replaced theories were empirically successful. Even if scientists had somehow failed to come up with new theories, the old theories would not have ceased to be successful. So success is one thing, replacement is another."[7] The science of any given time is the best empirical understanding achieved at that time, and, as such, it is to be trusted. With this premise—that the public ought to trust the view of the majority of scientists who are experts on the respective topic—let us now explore public attitudes toward science.

Varieties of Trust

The question might appear simple. Does the public trust science? If yes, does the public trust science as much as it should? Does the public trust science too little? Or too much? To answer these questions, we have to look at public attitudes toward science, mostly drawing on findings from opinion polls and occasionally on academic research on related topics. The aim here is to get a sense of how much the public trusts and distrusts science and what role uncertainty, or perceptions of uncertainty, plays in these attitudes. Unfortunately, polls like these are conducted systematically in only a few parts of the world, most often in the United States and Europe. Therefore, in this chapter, we focus on the United States; this might initially look like a case where we are looking for our lost keys where the light is better and not where we actually lost them, as in the old English joke. Nevertheless, the United States is one of the countries where cutting-edge scientific research is conducted and where, at the same time, strong anti-science movements are active, some of which will be discussed in subsequent chapters. For this reason, it is interesting to look in some detail into what people in the United States think about science.

The first question one might ask is how high or low is the public's confidence in scientists in the United States? An answer is found in the results of the General Social Survey (GSS), a project of an independent research organization, the National Opinion Research Center (NORC) at the University of Chicago. The public's confidence in the scientific community has been relatively stable for a period of 40 years between 1972 and 2012. Over this period, between 37% and 43% of participants have indicated that they have a great deal of confidence in the scientific community, whereas between about 38% and 51% of them indicated that they only had some confidence in it (with sample sizes for the surveys varying from 971 to 2,000 people). There are several ways to interpret these findings. On the one hand, more than 2 out of 3 participants indicated that they had at least some confidence in the scientific community, with less than 7% indicating that they hardly had any confidence in it at all. On the other hand, over these 40 years there were more people who only had some confidence in the scientific community than people who had a great deal of confidence in it (with the exception of 1974, 1976, and 1987). Since 1988, the number of people having only some confidence in the scientific community has been consistently more than those having a great deal of confidence in it.[8]

Is this a reason for concern? Perhaps. One can view the glass as half-full or as half-empty. More than 9 out of 10 people in the United States have at least some trust in scientists, but more than half of them only have *some* trust and not a great deal of it. The big question here is why this is the case. What makes people see scientists as undeserving of a great deal of trust? Their habits of mind? The nature of their endeavor? These questions are difficult to answer. It could be that people do not have a great deal of trust in science because they have heard stories of scientific misconduct or fraud. Thus, they may think that scientists are to be trusted only with caution. At the same time, it is interesting to note that during this same 40-year period, only the military has been considered more trustworthy than scientists. In contrast, confidence in Congress, the press, television, organized religion, and education has declined overall, and the level of trust in each of these institutions is lower than the overall trust in scientists.[9] Therefore, even if trust in science is not as high as one might like it to be, it is relatively high if one compares it to trust in other US institutions. Nevertheless, the fact that about half of the people in the United

States have only some trust in scientists is a reason for concern, especially given the anti-science movements in this country.

More recent surveys, after 2012, paint more or less the same picture. A survey conducted by Pew Research Center between May 10 and June 6, 2016, asked participants the following question: "How much confidence, if any, do you have in each of the following to act in the best interests of the public?" Among 3,014 participants who were asked this question about scientists, 21% answered that they had a great deal of confidence and 55% that they had a fair amount of confidence, whereas 18% stated that they did not have much confidence and 4% that they had no confidence at all. At first sight these more recent results seem encouraging: 76% of participants had at least a fair amount of trust in science. Within the same sample, 52% had a great deal or a fair amount of trust in religious leaders, 66% felt the same for public school principals and superintendents for grades K–12, and 41% felt the same for business leaders. Furthermore, trust was higher only for the military and medical doctors: among 4,563 participants asked the same question, 79% had a great deal or a fair amount of trust in the military and 84% felt the same for medical doctors. In contrast, only 27% of participants reported a great deal or a fair amount of trust for elected officials, and 38% felt the same for the news media.[10] Simply put, the situation remains the same: only about 4 out of 10 people in the United States have a great deal of trust in science.

The same survey also asked participants some interesting questions about climate change, one of the most hotly debated science topics in the United States. Among 1,534 participants, 48% agreed with the statement that "The Earth is getting warmer

mostly because of human activity such as burning fossil fuels," 31%
agreed with the statement "The Earth is getting warmer mostly
because of natural patterns in the Earth's environment," and 14%
agreed with the statement "There is no solid evidence that the
Earth is getting warmer." Interestingly, among these participants,
36% stated that they care a great deal about climate change and
38% stated they care some about it, whereas 18% stated that they
do not care much and 8% that they do not care at all about climate
change. However, when it comes to climate scientists, it seems that
the majority of participants thought that they should have a central role in making related political decisions: 67% thought that
climate scientists should have a major role in making decisions
about policy issues related to global climate change, whereas
56% thought that this should be the case for the general public,
53% for energy industry leaders, 45% for leaders from other nations, and 44% (!) for elected officials. The latter finding might
seem surprising. However, it is likely explained by the low level
of trust that people have in elected officials in general. When the
same participants were asked the question "How much, if at all, do
you trust each of the following groups to give full and accurate information about the causes of global climate change?," only 4% of
participants said that they had a lot of trust in elected officials and
only 25% said that they had some trust. This clearly indicates that
participants did not consider their elected officials trustworthy on
this particular topic. When asked the same question about climate
scientists, 39% stated that they trusted them a lot and another 39%
that they had some trust in them on this particular topic.[11] The
overall picture remains the same in the particular case of climate
science. The majority of the public has some trust in scientists, but

they are divided. Approximately half of the public trusts scientists a great deal, whereas the other half only has some trust in them. The question then is, why?

Whence the "Only Some" Trust in Scientists?

It should be obvious that holding views that go contrary to what contemporary science suggests, such as that global warming is not due to human activities, and having only some trust in scientists may be causally related. However, this cause–effect relation could go either way. On the one hand, the limited trust could be the cause of the different views: people may not accept the conclusions of scientists because they do not trust them in the first place (e.g., because they do not think highly of them as professionals). On the other hand, the different views could be the cause of the limited trust: people do not trust scientists because scientists' views conflict with their own (political or religious) views. To definitively figure out what is the case, one would have to investigate all parameters at the same time (trust in scientists, agreement with scientists' views, perception of scientists as professionals, etc.). Nevertheless, some conclusions can be drawn based on the data that are already available.

One interesting finding of the 2016 Pew Research Center survey on climate science, discussed earlier, was that the political views of participants had a major impact on their trust in scientists. Assuming a continuum of political views from being a conservative Republican, to being a moderate/liberal Republican, to being a moderate/conservative Democrat, to being a liberal Democrat, the same survey provided data supporting the conclusion that Republicans were overall more skeptical of the information, the understanding, and the research findings of climate scientists than were Democrats. For instance, it was found that 70% of liberal Democrats trusted climate scientists a lot to give full and accurate information about the causes of climate change, compared to only 15% of conservative Republicans. At the same time, about 54% of liberal Democrats stated that climate scientists understand the causes of climate change very well, which was the case for only 11% of conservative Republicans and 19% of moderate/liberal Republicans. Furthermore, about 55% of liberal Democrats stated that climate research reflects the best available evidence most of the time, whereas 39% of them stated that this was the case some of the time. In contrast, only 9% of conservative Republicans stated that climate research reflects the best available evidence most of the time, and 54% of them stated that this is the case some of the time. This clearly indicates a correlation between political views and trust toward scientists in the particular case of climate science. Nevertheless, 80% of liberal Democrats, 76% of moderate/conservative Democrats, 69% of moderate/liberal Republicans, and 48% of conservative Republicans stated that climate scientists should have a *major* role in policy decisions related to the climate.[12]

What is even more striking is that, contrary to what one might expect, knowledge about science does not seem to strongly impact opinions about climate issues. In other words, whether people know or do not know much about climate science in particular and science in general does not seem to be have an influence on how much they trust climate scientists. The 2016 Pew Research Center survey included nine questions about science topics, two of which were directly related to climate science. In those two cases, it was found that 68% of participants were aware that carbon dioxide is a gas created by burning fossil fuels, but only 27% of them were aware that nitrogen gas makes up most of the Earth's atmosphere. Overall, 22% of participants had a relatively high level of science knowledge, answering at least seven of the nine questions correctly. About 48% of participants had a medium level of knowledge, getting between three and six correct answers. Finally, 30% of them had a low level of science knowledge, getting no more than two answers correct. Overall, participants' general knowledge about science was found to be only modestly related to their views about climate issues; there was no strong, direct connection between participants' science knowledge and their beliefs about the causes of global warming, climate scientists, and the factors influencing climate research.[13]

A final interesting finding was that the particular influence of science knowledge on views about climate science varied based on political orientation. Normally, one should expect people's views about whether scientific understanding of climate change is accurate to be related to their knowledge of science, as people who know more about science should tend to perceive a consensus among climate scientists that human activities are responsible for climate change. Surprisingly, this was only the case for Democrats. Whereas Democrats with medium or high levels of scientific knowledge were more inclined to perceive a consensus among climate scientists than Democrats with low levels of scientific knowledge, there was no such difference among Republicans.

In particular, that the Earth is warming due to human activity was something accepted by 93% of Democrats with high science knowledge, 71% of Democrats with medium science knowledge, and 49% of Democrats with low science knowledge. This is a clear pattern indicating that the more knowledge of science one has the more likely it is for one to agree with climate scientists—but only if one is a Democrat. The respective numbers for Republicans were quite different: that the Earth is warming due to human activity was accepted by 23% of Republicans with high science knowledge, 25% of Republicans with medium science knowledge, and 19% of Republicans with low science knowledge.[14]

Before exploring further why Republicans think differently about this issue than Democrats, it is worth considering the findings of a Gallup poll on the same topic. Since 1997 and until March 1, 2018, Gallup has been asking participants whether they think that most scientists believe that global warming is occurring, that most scientists believe that global warming is not occurring, or that most scientists are unsure about whether global warming is occurring or not. The amount of people responding that most scientists believe that global warming is occurring was 48% in 1997 and 66% in 2018, albeit with some fluctuations. Less than 10% of respondents during the whole 1997–2018 period answered that most scientists believe that global warming is not occurring. Nevertheless, at the same time, a significant portion of the US public thought that most scientists are unsure about whether global warming is occurring or not. This has been declining overall from 39% in 1997 to 24% in 2018, again with some fluctuations. But about 1 in 4 of the 1,041 adults who participated in March 2018 seemed to misconstrue uncertainty in science. Now what is most interesting for our discussion here is that there are significant differences in these views depending on participants' political orientation. In particular, whereas 86% of Democrats thought that most scientists believe global warming is occurring, this was the case for only 42% of Republicans in 2018 (the respective amount in 2017 was the same for Democrats and 53% for Republicans). This implies that 58% of Republicans think either that scientists believe that global warming is not occurring or that scientists are unsure about this.[15] The perceived uncertainty among scientists could thus be the cause of low acceptance of the scientists' proposed causes of global warming.

Why would these people perceive scientists as being uncertain? There are several possible reasons, but one that has come to public attention recently is the question of the rigor and quality of contemporary science, to which we turn in the next chapter.

5 | Is Scientific Rigor Declining?

Is it unavoidable that most research findings are false, or can we improve the situation? A major problem is that it is impossible to know with 100% certainty what the truth is in any research question. In this regard, the pure "gold" standard is unattainable.

—*John P. A. Ioannidis*[1]

Can Scientific Research Be False?

In a highly cited article from which the epigraph to this chapter is drawn, epidemiologist John P. A. Ioannidis has argued that most published research findings are false. Hang on! Scientific research findings are false? What is going on here? Is Ioannidis an opponent of science? No, he is not. Rather, he is a very accomplished scholar: he is actually the "C. F. Rehnborg Chair in Disease Prevention at Stanford University, Professor of Medicine, Professor of Health Research and Policy, and Professor (by courtesy) of Biomedical Data Science at the School of Medicine; Professor (by courtesy) of Statistics at the School of Humanities and Sciences; co-Director, Meta-Research Innovation Center at Stanford; [and] Director of the PhD program in Epidemiology and Clinical Research."[2] Perhaps his university really likes him, and they exaggerate a bit? No. He is also one of the most highly cited researchers in the world with 188,067 citations of his published work as of our writing this.[3] Of course, as he himself writes on his Stanford University page: "Current citation rates suggest that I am among the 10 scientists worldwide who are currently the most commonly cited, perhaps also the currently most-cited physician. This probably only proves that citation metrics are highly unreliable, since I estimate that I have been rejected over 1,000 times in my life."[4] Much as we agree with him about citation metrics, the number of his citations definitely indicates that many scholars all over the world take his work seriously.

So why has this highly cited scientist argued that most published scientific research findings are false? Why have he and his colleagues also argued that many new proposed associations and effects are false or exaggerated[5] and that the translation of new knowledge into useful applications is often slow and perhaps inefficient,[6] as well as that most clinical research is not useful?[7] Is Ioannidis trying to take down science? No, not at all! Quite the contrary, Ioannidis has argued that science is very important and therefore governments should carefully consider science when making decisions about policy and regulations:

Our society will benefit from using the best available science for governmental regulation and policy. One can only applaud when governments want to support the best possible science, invest in it, find ways to reduce biases, and provide incentives that bolster transparency, reproducibility, and the application of best methods to address questions that matter. However, perceived perfection is not a characteristic of science, but of dogma. Even the strongest science may have imperfections. In using scientific information for decision-making, it is essential to examine evidence in its totality, recognize its relative strengths and weaknesses, and make the best judgment based on what is available.[8]

Ioannidis makes two very important points here: first, that perfection is a feature of dogma, not of science, and second that, despite imperfections, the best available scientific evidence should be considered for decision-making. So he argues that science is very important but also that we need to be cautious about which conclusions are drawn from scientific findings.

But why does science yield findings and conclusions that are wrong or false? Overall, there are two main reasons that this sometimes happens. The first reason is that there exist uncertainties inherent in science itself that have to do with the lack of crucial data or with the limitations of the methods and models used by scientists. The second reason has to do with the fact that scientists themselves are humans who may fail to overcome particular weaknesses inherent in human nature. There is a great variety of human weaknesses. Some are unconscious and relate to the limitations of our cognition and the resulting problems in interpreting the available data, as we saw in Chapter 1. Others may or may not be conscious and have to do with the tendency of people to pick out data that confirm their preconceived views (confirmation bias) or to talk about topics they are not really familiar with but about which they feel competent (stretch of expertise). And, finally, there are those situations in which scientists make arguments that go contrary to the available data (because of hype or financial interests) or even manipulate data to arrive at the results they wish to obtain (scientific misconduct). Obviously, these two kinds of reasons that lead to false scientific findings and conclusions are different. The first has to do with how science is done, whereas the second has to do with human nature. But before exploring possible solutions to these problems, it is necessary to understand the extent of these problems. The first kind of problem—that is, the uncertainties inherent in science—are explored in some detail in the five subsequent chapters on climate science (Chapter 6), vaccination (Chapter 7), human evolution (Chapter 8), genetic testing (Chapter 9), and forensic science (Chapter 10), and some broader conclusions about scientific uncertainty are drawn in Chapter 11. So, in this chapter, we focus on the second kind of problem, which is due to the inherent weaknesses of human nature.

Does Bias Exist in Science?

Bias does exist in science. As a result, even articles published in top science journals have been retracted. In 2010, Ivan Oransky, president of the Association of Health Care Journalists, and Adam Marcus, managing editor of *Gastroenterology & Endoscopy News* and *Anesthesiology News*, launched *Retraction Watch*, a blog that reports on retractions of scientific papers.[9] According to journalist Richard Harris, who has written a well-documented book about the problems with reproducibility and rigor in scientific research, *Retraction Watch* went from 40 retractions in 2001 to 400 retractions in 2010, and there have been about 500–600 retractions per year since then.[10] Obviously, something is wrong here. But what is it? Are we seeing fraud or flaw?

In 2009, Daniele Fanelli, at that time a researcher at INNOGEN and the Institute for the Study of Science, Technology & Innovation (ISSTI) of the University of Edinburgh and currently a Fellow in the Department of Methodology at the London School of Economics and Political Science, published an article with the title: "How Many Scientists Fabricate and Falsify Research? A Systematic Review and Meta-Analysis of Survey Data." This article was a meta-analysis of 18 studies that had directly asked scientists whether they had committed scientific misconduct or knew of a colleague who had. Scientific misconduct consists principally of (1) fabrication, such as inventing data; (2) falsification, which is the willful distortion of data or results; or (3) plagiarism, which is copying ideas, data, or words without attribution. Fanelli focused on fabrication and falsification and looked into the available studies up to that point. His main finding was that an average of 1.97% of scientists had admitted to have fabricated, falsified, or modified data or results at least once whereas 33.7% had admitted to other questionable research practices. When asked about the behavior of their colleagues, 14.12% of scientists reported falsification of data and 72% reported other questionable research practices. As shockingly high as these rates of misconduct are, Fanelli claimed that "Considering that these surveys ask sensitive questions and have other limitations, it appears likely that this is a conservative estimate of the true prevalence of scientific misconduct."[11]

These findings are certainly disconcerting. But one might still argue that the vast majority of scientists do not consciously fabricate their data. Not all scientists are honest, but most of them are. Unfortunately, even if we accept that the majority of scientists are honest and follow the ethical codes of research, there are still problems. Why? Because humans are involved in the conduct of science. The way science tends to be represented might make one forget this. We often read in scientific articles and books that "research has shown" or that "data indicate" or that "evidence suggests," etc. All these expressions are misleading, if not completely wrong. Through their research, scientists produce or obtain data that become evidence within a particular theoretical framework. Neither research, nor data, nor evidence "show," "indicate," or "suggest" anything on their own. They do not "speak for themselves." In contrast, everything in science is a matter of interpretation. Therefore, it would be more appropriate for scientists to use expressions such as "research has supported the conclusion that . . ." or "the conclusions drawn from the data obtained have been that . . ." or that "the available evidence has been interpreted as showing that. . . ." Indeed, these expressions might seem odd, and one might think of the expressions "research has shown" or "data indicate" or "evidence suggests" as a kind of shorthand in their place. But readers should be reminded that conclusions and interpretations can be subjective.

There are many examples of this kind in the history of science. Nonetheless, one might think that scientists in the past were simply less objective than those working today. Well, recent evidence might make you think otherwise. Raphael Silberzahn, assistant professor at IESE Business School in Barcelona, and Eric L. Uhlmann, associate professor at INSEAD in Singapore, conducted an interesting experiment. They recruited 29 teams of researchers and asked them to answer the following question: Are football (soccer) referees more likely to give red cards to players with dark skin than to players with light skin? The researchers who agreed to participate in this experiment held different opinions about whether such an effect existed. They were all given the same large dataset from four major football leagues, which included referee calls, counts of how often referees encountered each player, and player demographics including the players' position on their teams, height, weight, and skin color. Each of the 29 teams of researchers developed its own method of analysis. All methods were sent back to all researchers, anonymized and without revealing results, asking them to rate the validity of each method and to provide in-depth feedback on three of them. In the end, the teams were given the opportunity to revise their method based on the feedback received. Twenty of the 29 teams found a statistically significant correlation between skin color and red cards. Dark-skinned players were found to be 1.3 times more likely than light-skinned players to receive red cards. However, the findings varied enormously, from a slight (and nonsignificant) tendency for referees to give more red cards to light-skinned players to a strong tendency of giving more red cards to dark-skinned players. Thus, an important conclusion was reached: "Any single team's results are strongly influenced by subjective choices during the analysis phase. Had any one of these 29 analyses come out as a single peer-reviewed publication, the conclusion could have ranged from no race bias in referee decisions to a huge bias."[12] Figure 5.1, used in that article, provides a clear take-home message. Subjectivity is there, and we should keep it in mind. As the authors noted, "taking any single analysis too seriously could be a mistake, yet this is encouraged by our current system of scientific publishing and media coverage."

FIGURE 5.1 How the same data can be interpreted differently by different scientists.

What are we to make of this? Subjectivity in the interpretation of data is absolutely normal and should not affect our overall trust in science. Rather, it should make scientists reconsider how they themselves evaluate research and how the rest of us understand it. In principle, scientific research should be reliable and valid when it comes to measurements and data collection. *Reliability* is about obtaining the same result every time one uses the same instrument to make the same measurement, whereas *validity* is about measuring exactly what the instrument is intended to measure. To make this clearer think of a ruler and a measuring rope. No matter how many times we measure a certain length with the ruler, we will likely acquire the same result because the ruler is rigid. This is not the case for the rope, which might be stretched. In this sense, the measurements made with the ruler are more reliable than those made with the rope. In the same sense, the ruler may be graded more accurately and thus better correspond to actual units (e.g., centimeters) than the rope. As a result, the measurements made with the ruler are more valid than those made with the rope. Therefore, when using a ruler, one might be confident that if one measures 1 cm, it is indeed 1 cm, and one will make the same measurement no matter how many times one uses the ruler.

Returning to science, if the published scientific research is solid, then one should be able to use the same methods that other researchers previously used and arrive at the same results. This is what *reproducibility* in science is about. An online survey conducted by the prestigious science journal *Nature* asked 1,576 researchers about their thoughts on the reproducibility of research. Among these, 52% thought that there is a significant reproducibility crisis in science, 38% thought that there is a slight crisis, and only 3% thought there is no crisis, whereas 7% replied that they did not know. More than 60% of participants in all domains reported that they had failed to reproduce someone else's research, and more than 40% of participants in all domains reported that they had failed to reproduce *their own* research. Nevertheless, fewer than 20% of them had ever been contacted by another researcher unable to reproduce their work, and 73% of them said that they think that at least half of the papers in their field can be trusted. Interestingly, only 24% of them said that they had been able to publish a successful replication and 13% that they had published a failed replication. Among the actual attempts to reproduce research findings, only 11% (6 out of 53 published studies) for cancer biology[13] and 36% (36 out of 100 published studies) for psychological science have been successful.[14] What is the reason for the reproducibility crisis? More than 60% of the scientists said that selective reporting, scientists not reporting all the details of their experiments as they ought to, and pressure to publish in order to get tenure or grants always or often contributed to the problem by making scientists conduct their research less diligently than they should.[15] This sounds like a reasonable explanation, but is it?

Scientific Rigor Is Alive and Well, Despite a Few Exceptions

Are scientists striving to publish whatever they can, failing to report the details of their experiments, and perhaps failing to do science rigorously? Is this why we encounter such low research reproducibility? Let us look at these issues one by one. The first point to note is that, contrary to what one might expect, it is not always clear what "research reproducibility" is. This term has not been consistently used across all fields; at the same time, it is often not easy to distinguish between reproducibility and terms such as "replicability" and "repeatability" despite their differences. It is therefore important to first clarify what we are talking about. A suggestion that has been made is to distinguish among three types of reproducibility: methods reproducibility, results reproducibility, and inferential reproducibility. *Methods reproducibility* refers to the use of the same methods as previous research; that is, reimplementing the same methods as an older study as exactly as possible. *Results reproducibility* refers to reproducing the same results as an older study in a new study after implementing the same methods (what is often described as replication). Finally, *inferential reproducibility* refers to drawing the same conclusions as a previous study; this should be distinguished from results reproducibility because, as we saw earlier, it is not necessary that all scientists will draw the same conclusions from the same data. Overall, reproducibility refers to using the same methods to produce the same results from which the same conclusions are drawn.[16] Whenever this happens, we can be very confident of the strength of the respective knowledge produced.

Let us now turn to the major causes of scientists' perception of low reproducibility in scientific research. The primary cause identified by scientists who perceived a reproducibility crisis was the bad reporting of results. Whether this is actually the primary cause or not is difficult to estimate and depends on the field. In one case, the editors of *The Journal of Cell Biology* reported that, among more than 4,000 papers that had been accepted for publication, about 15% contained inappropriate presentation of data. Most of these were subsequently corrected by the authors, leaving only 1% of the papers unable to be published due to serious discrepancies.[17] It could be the case that the problem is not really due to bad reporting of results but due to other reasons. It is important to note that whereas methods reproducibility might be achievable all the time, this may not be the case for results reproducibility and inferential reproducibility for purely practical reasons, as one study in that same journal showed.[18] This is, of course, something that can be effectively addressed by journals such as, for example, *Nature* and *Science* striving for more transparency in the presentation of methods and data.[19]

Is the problem then that scientists rush to publish in a "publish or perish" spirit, where the more one publishes, the more productive one is considered, and therefore the more likely one is to get tenure or funding? According to a recent estimation, some scientists publish about one paper every 5 days. The researchers who looked into this searched Scopus for authors who had published more than 72 papers in any one calendar year between 2000 and 2016 (which is the equivalent of one paper every 5 days). What is more interesting is that this has not been achieved by a few exceptional individuals, but by more than 9,000 people! The majority of these authors were in physics (7,888). It was noted that in high-energy and particle physics research is conducted by large international teams with more than 1,000 members who are listed as authors because of their membership, not for writing or revising the papers. This, of course, raises many questions about what authorship requires and entails.[20] However, in another study, Daniele Fanelli also found that scientists may not publish more now than in the past. In particular, he analyzed 760,323 papers published between 1900 and 2013, by 41,427 authors who had published two or more papers within the first 15 years of their career in any of the disciplines covered by the Thomson Reuters Web of Science database. Whereas the numbers of papers published by early-career researchers and the total number of authors has increased in recent times, individual productivity was not found to have increased.[21] These findings do not support the assumption that there is a higher pressure on early-career researchers to publish now than in the past. But there may still be pressure to develop collaborations with colleagues, not only because they are necessary for doing science but also for expanding one's publication record. Important questions thus arise: How should a scientist who publishes by virtue of being a member in a consortium without ever writing (or perhaps even reading) the papers under his or her name be compared to someone else who belongs to small research group, all the members of which actually contribute to publishing the paper? How should a scientist co-authoring numerous papers, even in high-profile journals, with numerous other authors be judged if he or she has only written very few papers alone?

Another important issue relates not only to how many papers are published but also to their content. As already mentioned, there are not many replications studies published because people strive to produce something "novel," which is perceived to be more valuable than replicating previous works, both by journal editors and by other scientists. It is indeed the case that in many research fields papers are more likely to be accepted for publication, especially in high-profile journals, and to be cited if they report "positive" results (i.e., results that overall support the hypothesis tested by the researchers). This can be a bias both of the authors who would like to see their hypothesis confirmed and of the journal editors and reviewers who might feel that "positive" findings are more interesting to researchers in the field. As comedian John Oliver nicely put it during the episode "Scientific Studies" of his show *Last Week Tonight*, "to get published, it helps to have results that seem new and striking, 'cause scientists know that nobody is publishing a study that says "nothing up with açai berries."[22] Daniele Fanelli tried to test the hypothesis that there is a connection between pressures to publish and bias against "negative" results in the scientific literature by analyzing a random sample of 1,316 papers that were published between 2000 and 2007 with a corresponding author based in the United States. He explained that the geographical origin of the corresponding author should not be expected to have an effect on the quantity of publications per capita. Nevertheless, he considered the results of his analysis as supporting the hypothesis that competitive academic environments increase both the productivity of researchers and their bias against "negative" results.[23] This is certainly a cause for concern, especially given the value of replication studies and the low amount of reproducibility.

Science Should Be Trusted Because of Its Self-Correcting Nature

Even if data are not badly reported and even if scientists do not rush to publish, problems exist. The most indicative illustrations of these are retracted papers, already discussed in the previous section. These are papers that were initially published—even in top journals—and that were retracted because they were found to have flaws or signs of fraud. Once again Fanelli looked in some detail at this situation by examining 2,294 articles that included the term "retraction" in their title. He found that, in recent years, there has been an increase in the number of journals issuing retractions, but not in the number of retractions per journal. At the same time, the number of queries and allegations made to the US Office of Research Integrity has increased, but this has not been the case for the number of cases of misconduct. This suggests that the growing number of retractions is a sign that journal editors are getting better at identifying and retracting flawed and fraudulent papers, rather than an increase in scientific misconduct. If this is so, it is a positive sign that science is self-correcting.[24]

Self-correction is a very important feature of science that can be illustrated with a simple example. A person, let's call him Opt (for optimist), never goes to the doctor and feels healthy. Opt will likely have nothing to worry about until later in life—perhaps until it is too late. Yet, until that time, Opt can claim that he is healthy and that he has nothing to worry about. A friend of his, Caut (for cautious), is more proactive, perhaps even a bit of a hypochondriac. He regularly visits the doctor from an early age, and, as soon as he realizes that his sugar and lipids levels, or his blood pressure, are above the normal range, he starts losing weight and exercising more than before. Caut starts feeling awkward because he cannot eat and drink as much as his friend Opt, and he has to walk to work in order to exercise. Opt even makes fun of him, calling him an imaginary patient. Should Caut feel embarrassed? No! Caut is doing what he can to lead a better life in the long term;

he is self-correcting his life habits. Opt has left things to chance and does not worry if he does harm to his body. Science is like Caut, not Opt; it is a self-correcting enterprise that aims for the best possible condition. It's not embarrassed to correct itself, unlike other enterprises that are either unaware of their problems or prefer to live in eternal ignorance. It this self-awareness and self-correcting aspect of science that provides a good reason to trust it despite its failings. As it was nicely put recently:

> The discovery that an experiment does not replicate is not a lack of success but an opportunity. Many of the current concerns about reproducibility overlook the dynamic, iterative nature of the process of discovery where discordant results are essential to producing more integrated accounts and (eventually) translation. A failure to reproduce is only the first step in scientific inquiry. In many ways, how science responds to these failures is what determines whether it succeeds.[25]

PART II

Uncertainties in Science

SOME CASE STUDIES

6 | Uncertainties in Climate Science

> The uncertainties in the IPCC 2013 report are slightly
> higher than those in the 2007 report and this is because of
> our greater understanding of the processes and our ability to
> quantify that knowledge.
>
> —*Mark Maslin*[1]

What Is Climate Change?

This chapter is the first in a series of case studies examining
uncertainties present in specific sciences. What is particularly interesting is that confusion about the nature of the uncertainties involved in these sciences, which arise because of misinformation or
simple misunderstanding, can lead to different sorts of problems.
In some cases, uncertainties are given too much emphasis and are
taken by a number of people to discredit the veracity of the respective science—we see this in the case of climate change, discussed
in this chapter, as well as in the cases of vaccination and human
evolution. In other cases, uncertainties are underemphasized,
leading people to ignore them and as a result have unreasonable
expectations of the respective science—this is common when
it comes to genetic testing and forensic science. In this chapter
and the subsequent ones, we explore the uncertainties inherent
in these sciences with the aim of avoiding the common pitfalls

to which misunderstanding uncertainty gives rise. We argue that different kinds of uncertainties are inherent in these sciences and that people need to be aware of them in order to understand what these sciences can and cannot achieve. Let us begin here with climate change.

In a May 2014 interview, then US President Barack Obama claimed that climate change "is not some distant problem of the future. This is a problem affecting Americans right now."[2] This is not just an American problem, of course. Climate change is something that affects everyone in the world. But what exactly is it? Is climate change the same thing as global warming? No. Global warming is an important part of climate change, but it is not the whole story. Global warming is "the increase in Earth's average surface temperature due to rising levels of greenhouse gases," whereas climate change is "a long-term change in Earth's climate, or of a region on Earth."[3] US President Donald Trump was correct when he tweeted that scientists chose to reframe discussions in terms of "climate change" rather than "global warming" because the latter term was not working.[4] However, the reason for this was not, as he implied, that global warming is not occurring. Instead, the reason for the change in terminology was that there is more to the problem than simply global warming. Temperature change alone is not the most severe effect of the changing climate. According to the Intergovernmental Panel on Climate Change (IPCC), a group of hundreds of climate scientists, "it is *very likely* that . . . extreme precipitation events will become more intense and frequent in many regions. The ocean will continue to warm and acidify, and global mean sea level to rise."[5] Hence, the term "climate change" better expresses the range of threats we face than does the term "global warming."

None of this is to say that global warming has not been occurring. Indeed, it has. Global temperature records began to be kept in the mid-1800s. Since then, the 1980s was the hottest decade on record until the 1990s, which was hotter than the 1980s. The period from 2000 to 2009 was even hotter than the 1990s, and 2010–2014 was even hotter than that![6] This trend of a continuous increase of the average temperature is ongoing. The year 2014 was the hottest year on record until 2015 set a new record. And, 2016 ended up being even hotter than 2015.[7] As climate scientist Joseph Romm has put it "the world's leading scientists and governments have stated flatly, 'Warming of the climate system is unequivocal'; it is a 'settled fact.'"[8]

Now one might question, as people like President Trump have, why should we think that global warming and climate change are really happening when we are experiencing unseasonably cool weather at times in the United States and elsewhere? After all, why should we have intense winter storms and unseasonably cold weather if the climate is changing and the global temperatures are rising? This fairly common misunderstanding of climate change rests on a confusion between "climate" and "weather". *Weather* is the specific atmospheric conditions (temperature, precipitation, etc.) in a particular place at a particular time. *Climate*, on the other hand, is the average weather conditions over a long period of time. So the weather in New York City is the specific conditions one finds there on a particular day at a particular time; the climate of New York City is an average of weather conditions over a long period of time (often decades).[9] Hence, when we talk of climate change, we are talking about the average of weather conditions of the various regions of the world over long periods of time.

Similarly, global warming concerns the average of temperatures from all over the world over a long period of time. Consequently, the fact that it is colder than normal in a particular city during a particular year, which was of concern to President Trump (see the epigraph of Chapter 3), cannot cast doubt on the reality of global warming or climate change. An analogy will help illustrate this point. When a professor says that the exam scores on the second exam were higher than those on the first, this can be true even if individual students did not improve. In fact, it can be true even if some students performed worse on the second exam than on the first. All that matters is that the *average* of *all* test scores was higher for the second exam than for the first. The same is true of global temperatures. The fact that one particular region of the world experiences lower average temperatures than the previous year does not mean that the global average temperature has not increased. Thus, pointing out an unseasonably cool temperature in a particular place does not demonstrate anything about climate change or global warming because these are both concerned with averages of the entire world over long periods of time.

What Is the Evidence for Climate Change?

It is worth briefly considering some of the evidence for climate change because doing this can help us better appreciate where the uncertainties lie. One of the most obvious sources of evidence for climate change is the rising of average global temperatures. We noted earlier that each of the past few years has broken the previous year's record for average temperature. "Warming of the climate is unequivocal"—in fact, "the period from 1983 to 2012 was *likely* the warmest 30-year period of the last 1,400 years in the Northern Hemisphere."[10] These temperature records are the result of careful measurements made with thermometers and other instruments. And they have been corroborated by looking at things like tree rings and ice cores.[11] The width of tree rings provides information about the climate conditions experienced during the year that the ring was formed. For example, wider rings tend to signal warmer years whereas narrower rings are a sign of colder conditions.[12] Ice cores provide information about atmospheric conditions in the past because small air pockets get trapped under snow as it falls and become part of the ice core. These air pockets are samples of what the atmospheric conditions were like when that particular layer of the ice core was formed.[13] The evidence from tree rings and ice cores is independent of the temperature measurements taken; therefore there is independent evidence converging in support of increasing global temperatures, and this is very important to note.

In addition to rising temperatures there is also the fact that human emissions of greenhouse gases are the highest they have ever been.[14] The primary greenhouse gas that humans are producing is carbon dioxide (CO_2). At around the beginning of the Industrial Revolution (about 250 years ago) CO_2 levels in the atmosphere were about 280 parts per million (ppm). They are now more than 400 ppm. Our emissions of CO_2 today are *six times higher* than they were in 1950![15] Why does this matter? Because the best explanation for the rapid increase in global temperatures is that it is the result of higher levels of greenhouse gases in the atmosphere. Greenhouse gases in the atmosphere trap heat that would otherwise escape the planet. As a result, the more molecules of such gases there are in the atmosphere, the more heat is retained in the atmosphere rather than escaping into space. Long story short, "it is *extremely likely* that more than half of the observed increase in global average surface temperature from 1951 to 2010 was caused by the anthropogenic increase in GHG [greenhouse gas] concentrations and other anthropogenic forces together."[16] Global warming is in large part due to human activities.

There seem to be two main causes of this problem. The first cause is the burning of fossil fuels, which is the main source of the CO_2 released into the atmosphere. The second major cause of the problem are changes in land use. Cutting down forests in order to use the land for agriculture or for building cities has diminished the areas that are covered by plants capable of absorbing CO_2 during photosynthesis. What is perhaps most troubling is that the situation is likely to get worse. It is estimated that between 2015 and 2044 humans will release about half a trillion tons of CO_2 into the atmosphere, which is equal to the amount of CO_2 released between 1750 and 2015. In other words, we are currently capable of producing within the next 30 years as much CO_2 as we did in the past 265 years. The reason for this is that countries such as China, India, and Brazil are currently increasing their emissions at a very fast pace.[17] We therefore produce a lot more CO_2 than before while having significantly fewer plants to absorb it. We have done about as poorly as we possibly could.

This is not the whole story though. There is other evidence of climate change in addition to direct measurements of rising temperatures and greenhouse gas emissions. There are various phenomena we should expect to observe if climate change were occurring. We should expect both the atmosphere and the ocean temperatures to rise. We should expect the amounts of snow and ice, particularly at the poles, to diminish. And we should also expect the sea level to rise. All these have been observed.[18] We should expect various weather effects as well. As the temperature of the ocean rises, more of its water will evaporate, leading to increased levels of humidity. Such increased levels of humidity have also been observed. As a result of the increase of water vapor in the air, we would also expect to see intense rainfalls occurring with greater frequency. We have seen this, too.[19] All in all, we are currently observing exactly what we would expect to see if climate change were occurring. This is powerful evidence in support of the existence of climate change.

This cursory examination of the evidence for climate change is sufficient to demonstrate that there is strong evidence from a variety of sources for climate change. Does this evidence make climate change certain? No. It is possible (in the broad sense of the term) that the evidence for climate change is misleading or that it has been misinterpreted. Nevertheless, the odds of either of these are miniscule. To put the strength of evidence for climate change in perspective, Romm has pointed out that scientists are as sure that climate change is real and "that humans are responsible for [the] most recent climate change as they are that cigarettes are harmful to human health."[20] Are they certain? No, but quite sure nevertheless!

Uncertainties Inherent in Climate Science

Although the evidence in support of climate change and the role that humans are playing in it is strong, uncertainties remain. This is not something that climate scientists deny. In fact, the IPCC is explicit about the known uncertainties surrounding the statements that it makes. For example, in its 2014 *Synthesis Report,* the IPCC labeled each of its claims from *virtually certain* (99–100%) to *extremely unlikely* (0–5%).[21] Hence, even the best grounded of the claims that the IPCC makes is not certain—even *virtually certain* statements are held to be somewhere between 99% and 100%, so they all admit of some uncertainty. As we will also see in later chapters, this is not something that we should find surprising because all science is uncertain. That being said, it is important to be clear about why uncertainties in climate change arise and where such uncertainties lie.

The primary sources of uncertainty when it comes to climate change are complexity and access to evidence. Let us start with the latter. Although we have very good evidence in support of climate change and humans' influence on it, we do not have precise temperature measurements prior to the 1800s. As a result, we are forced to estimate past temperatures as well as the nature of the global climate in the past on the basis of indirect evidence. In addition to this limitation with respect to evidence there is also the enormous complexity of the global climate, which is the result of numerous interactions about which many details are not well understood.[22] These uncertainties about the details of climate interactions are at the heart of disagreements concerning climate change among scientists. But none of these disagreements is about whether climate change is occurring or whether it is influenced by human actions. Rather, the disagreements concern how quickly and by what mechanisms climate change is occurring.[23] In fact, despite the uncertainties concerning the exact details of the mechanics of climate change, the reality of climate change is taken as so well established that the focus of scientists is on predicting how rapidly Earth will warm and the climate effects that we will see as a result—because they take it for granted that this will happen.[24]

Returning to the uncertainties, some of these arise from the fact that the global climate is affected by factors that are very difficult to predict, such as what exactly our future amounts of greenhouse gas emissions will be. The climate is also affected by natural phenomena such as the activity of volcanoes and the sun, which can be nearly (if not entirely) unpredictable. Other uncertainties stem from our use of scientific models for predicting climate change. Any scientific model is limited in the sense that it cannot fully represent every aspect of a target system (if it did, it would simply be the system itself and not its model). For example, a map represents some aspects of an area, such as where streets and towns are, but by no means every aspect of each street and town. This feature of models is especially true when the target system is something as complex as the global climate. Consequently, the models used to predict climate change have to make a number of simplifying assumptions and estimations. Each one of these assumptions and estimations introduces uncertainty because it is impossible to be aware of all the relevant factors in every detail. Furthermore, when scientists (or anyone else) construct a model designed to make predictions about climate change, they have to make judgments about how to weight certain assumptions and how best to make estimations. These judgments are affected by the social values of scientists, such as whether to weight the possibility of really bad outcomes more heavily than the possibility of other outcomes.[25] These judgments do not make predictions inaccurate or accurate;

they only affect the predictive aims of scientists constructing climate models. However, different predictive aims lead to different models. And different models are apt to make different predictions. This is why we face uncertainty when it comes to which predictions we should accept.

Can we mitigate the uncertainty that arises from having many distinct possible models for the same situation? Plausibly we can, at least to some degree. One way to do this is through what is known as *robustness analysis*. The idea here is that when considering predictions about the global climate, what we should do is look at the predictions of several different models. Predictions that hold "across a range of models with different assumptions, parameters, or types of representations" are robust.[26] The more robust a prediction, the more trustworthy it is. Of course, this will not fully mitigate the uncertainty that arises from having different models making climate predictions, but it holds promise for helping us reduce that uncertainty. And, of course, no model is perfect. Even our best models are going to yield mistakes. For example, the IPCC's 2001 and 2007 predictions for global temperature increase have turned out to be accurate, but the IPCC's predictions concerning the rate of sea level rise have been off. Unfortunately, the sea level has risen faster than the IPCC predicted.[27] While the use of techniques such as robustness analysis may help mitigate uncertainty when it comes to predicting climate effects and climate change, they cannot eliminate it.

As climate scientist Mark Maslin has explained, models allow for a big range of possible outcomes. The question then becomes what should be considered the best outcome? In a series of 10,000 simulations using a specific model, the average results matched the projections of the IPCC, but the extreme results were found to be equally probable. How should one choose what outcome to expect on this basis? Worse than this, when the results of particular models are used in higher resolution models expected to provide a better understanding, the range of the possible outcomes can become even larger. In some cases, the results of the original model and the high-resolution model are contradictory. Maslin has described this situation as the "cascade of uncertainty," where getting at a finer level of detail produces more uncertainty.[28]

Has the uncertainty in predicting climate effects and in understanding how exactly and how quickly climate change is occurring caused a lack of consensus among scientists concerning the reality of climate change and human influence upon it? No. Contrary to what many people believe (a recent survey found that 34% of people in the United States "don't know" how much consensus there is among scientists concerning climate change, and more than 50% underestimate the degree of consensus[29]), there is not a lack of consensus among scientists on these issues. Historian of science Naomi Oreskes analyzed the abstracts of 928 peer-reviewed scientific journal articles published between 1993 and 2003, which were listed in the Institute for Scientific Information (ISI) database with the keyword "climate change." She found that none of the papers disagreed with the consensus position that climate change is occurring and that it is being strongly influenced by human actions.[30] More recently, John Cook, professor at George Mason University's Center for Climate Change Communication, and his colleagues conducted a similar analysis in which they examined the abstracts of 11,944 peer-reviewed scientific journal articles published between 1991 and 2011 with topics on "global climate change" or "global warming." They found that when it comes to the consensus position that humans are causing global warming and climate change, it was endorsed in 97.1% of the analyzed articles whereas it was rejected in only 0.7% of them. In other words, the number of papers rejecting the consensus position was "a vanishingly small proportion of the published research."[31] Not surprisingly, the IPCC is not the only organization to have claimed that humans are causing global warming and climate change; this is the case for all major scientific bodies in the United States whose expertise bears directly on the matter.[32] The fact is that "virtually all professional climate scientists agree on the reality of human induced climate change."[33]

How Should We Deal with the Uncertainties in Climate Science?

Given the widespread consensus among climate scientists it may seem a bit puzzling that a third of the people in the United States are not aware of it and more than half underestimate how large the consensus is. There are likely a number of reasons for this confusion. One reason stems from the fact that "virtually all" is not all. There are some climate scientists who deny that human actions have a major causal influence on the global climate. On its own this would not explain the misperception of how strong the consensus is. You might think that scientists would only disagree for good reasons; however, they are human just like the rest of us. That means that even if they act on purely unselfish reasons and are guided solely by a desire to discover the truth, they still make mistakes. So, some of the scientists who disagree with the consensus might simply be making well-intentioned errors. Furthermore, in general, scientists are subject to the same biases and conflicts of interest as anyone else. The social structures of science may help to weed some of this out, but it cannot ensure that everyone operates only under the guidance of pure motives. As Naomi Oreskes's and Erik M. Conway's excellent book, *Merchants of Doubt*, made clear, some scientists have misled the public about important issues such as the dangers of tobacco smoke and climate change because of political leanings or/and financial incentives.[34]

Another reason that the consensus may be misperceived by the public comes from how the issue of climate change is treated by the media. Oreskes correctly pointed out that "the mass media have paid a great deal of attention to a handful of dissenters in a manner that is greatly disproportionate with their representation in the scientific community."[35] Presumably, this often occurs because the media has a desire to be perceived as fair and unbiased. "Fair and unbiased" is many times interpreted as reporting on both sides of an issue. So when news programs have a climate scientist speak about climate change they will often find another scientist who denies the consensus to offer an opposing viewpoint. In itself, this is not a bad thing. However, it can be deeply misleading to the general public watching the program. The average viewer sees two scientists disagreeing about an issue. They typically are not told that the scientist speaking in favor of the consensus view about human-caused climate change is representative of thousands of other scientists, whereas the dissenting scientist is representative of just a handful. This sort of disproportionate representation is very problematic, and it is apt to confuse the average viewer about the degree of consensus there is on this issue.[36]

None of this is to say that the consensus view *must* be correct. However, the evidence in support of climate change is solid, and there is almost unanimous agreement on the issue from the experts. The rational thing to believe is that climate change is real and that humans are causing it. Uncertainty about the details of how quickly climate change is occurring and by what mechanisms does not undermine this. We have ways of mitigating the uncertainty present in climate models to some degree by making use of techniques like robustness analysis. Nevertheless, this is not perfect; uncertainties remain. At the end of the day, the best we can do is follow the evidence and act accordingly. Despite the uncertainties inherent in climate science, we have strong evidence that climate change is happening largely because of human activities.

7 | Uncertainties in Vaccination

[A] choice not to get a vaccine is not a risk-free choice. It's just a choice to take a different risk.

—*Paul Offit*[1]

What Do We Know About Vaccination?

Vaccination is a very common practice in developed countries, one that saves millions of lives. Starting at very young ages, vaccines provide us with the means necessary to deal with pathogens and avoid the disease symptoms that these typically bring about. Pathogens usually come in the form of microbes (such as viruses and bacteria) that enter the human body and disrupt its physiology, causing a variety of symptoms and sometimes death. Vaccines help us avoid these problems by switching on our immune system to fight these microbes (immunization). Why is this necessary? Whereas our immune system can recognize and destroy almost every possible microbe that is foreign to it, it can only do so *after* its first encounter with the microbe. Initially, our immune system has small amounts of highly specialized white blood cells capable of killing each microbe. As a result, during the first encounter with a microbe, we cannot always successfully deal with it, which leads us to get sick—or even die. But once we have gotten sick and recovered there are many more white blood cells of the type that destroys the particular kind of microbe. Consequently, the next time that microbes of this kind get in our body, they are likely to all be destroyed fairly quickly. This is *naturally acquired immunity* to that particular microbe. However, this does not mean that we are immune to the disease caused by that microbe because different microbes can cause the same disease.[2] This is why we may have the flu every year. There are many different kinds of (related) viruses that cause the flu, and immunity to one kind does not amount to immunity to all others.

Vaccination helps overcome the limitation that we can only reap the benefits of a naturally acquired immunity after having been infected by some microbe. Vaccines can help us deal effectively with a microbe the very first time we encounter it because they activate our immune system so that the first time it encounters a microbe, it is equipped as if it had already dealt with the microbe before. How? Vaccines actually contain very small amounts of weak or dead microbes or their inactivated toxic molecules that are called *toxins*. When these particles or molecules are injected into our bloodstream, the appropriate white blood cells (which are already there) "recognize" them and proliferate. Because the microbes are weak or dead and because the toxins are inactivated, usually no disease develops. However, the proliferation of the white blood cells results in an increased number of them circulating in our bloodstream. As a result, if at some later point in time we are infected by the microbe that the vaccine targets, our immune system will effectively react to it. Vaccines thus provide *artificially acquired immunity*; our immune system is switched on artificially, preparing it ahead of time to fight the particular microbes and thus eliminating the risks of suffering and dying from a certain disease.

The available evidence shows that vaccines have contributed to a very significant decrease in a variety of infectious diseases. Vaccinations for nine vaccine-preventable diseases in the United States have resulted in more than a 90% reduction in their instances. And, in some cases, diseases have been almost or completely eliminated, such as smallpox and polio (Table 7.1).[3]

It is very important to realize that vaccines not only confer immunity at the level of the individual, but they also confer immunity at the community level by offering protection for unvaccinated people as well—as long as the vast majority of people are vaccinated. Let us explain. When all or most members of a community are vaccinated, then the possibility of an epidemic is limited. The probability that a person carrying a pathogen will encounter an unvaccinated member of the community, who would be more susceptible to developing the disease, is low. Most people who an infected person will encounter will be vaccinated and thus will be unlikely to have the pathogen proliferate inside them, and therefore they will be unlikely to transmit the pathogen to others. The vaccinated majority can thus protect the unvaccinated minority, resulting in what is called *herd immunity*. However, if there is a significant number of unvaccinated people within the community, then there is a higher probability that a person carrying a pathogen will encounter someone who lacks immunity, and so the probability of the pathogen being spread to others rises. In this case then, the unvaccinated people can become the means for the transmission of the pathogen and eventually for the outbreak of an epidemic. Therefore, one's decision not to be vaccinated is not a decision that affects one's own health only; rather, it is a decision that can have consequences

TABLE 7.1 Comparison of Representative Twentieth-Century Annual Morbidity Due to Infectious (Vaccine-Preventable) Diseases and 2016 Reported Cases

Disease	Twentieth-century annual morbidity	2016 reported cases	Percent decrease (%)
Smallpox	29,005	0	100
Diphtheria	21,053	0	100
Measles	530,217	69	>99
Mumps	162,344	5,311	97
Pertussis	200,752	15,737	92
Polio (paralytic)	16,316	0	100
Rubella	47,745	5	>99
Congenital rubella syndrome	152	1	99
Tetanus	580	33	94
Haemophilus influenzae	20,000	22*	>99

*Haemophilus influenzae type b (Hib) <5 years of age.

From Orenstein, Walter A., and Rafi Ahmed. "Simply Put: Vaccination Saves Lives," *Proceedings of the National Academies of Science* (2017): 4031–4033, p. 4031.

for the whole community (in some cases, these can be tragic consequences for those who cannot receive vaccines for medical reasons). This is very important to keep in mind because the more contagious a disease is, the higher the immunization rate of a community has to be in order for herd immunity to provide a safeguard against epidemics. For instance, for highly contagious diseases such as measles or pertussis the immunization rate has to reach about 95% of a community for there to be herd immunity, whereas for a less contagious disease such as mumps or rubella an 85% immunization rate can be sufficient. But one should also keep in mind that no vaccine is 100% effective, so the number of immune people will always be lower than the number of those vaccinated.[4]

Despite the obvious benefits of vaccination, some people worry about the effect of various ingredients that vaccines contain in order to enhance their impact. For instance, some vaccines contain aluminum as adjuvant; others contain antibiotics because these are used during vaccine production to prevent bacterial contamination. In some other cases preservatives are used in vaccines to prevent the growth of bacteria or fungi that may be introduced into the vaccine during its use (e.g., when the vaccine is administered by way of a multidose vial, the repeated needle punctures into the vial might carry microbes into the vaccine solution). Some vaccines, such as the live measles and mumps vaccine, may also contain egg protein because they are grown on cultured chick embryo fibroblasts. The concentrations of these proteins are very low and unlikely to trigger any allergic reactions, which is why they are considered safe overall for children with egg allergy. Although one might be concerned about these ingredients of vaccines, their possible impact is regularly tested before the vaccines are licensed, and they continue to be tested even after they have been licensed. The available evidence so far has not shown that these ingredients cause significant problems to the extent that they should not be used. For instance, the quantities of mercury, aluminum, formaldehyde, human serum albumin, antibiotics, and yeast proteins in vaccines have not been found to be harmful in humans or experimental animals. And although gelatin and egg proteins are contained in vaccines in quantities that might induce rare instances of severe hypersensitivity reactions, these vaccines are not considered to be problematic.[5]

All in all, vaccines are a safe and effective means for protecting ourselves from infectious diseases. However, as with all science, we cannot be 100% sure of either their effectiveness or their safety. Vaccination will not have a 100% success rate because some individuals will not have the anticipated immune response and might even develop the infectious disease for which they were vaccinated. Vaccination will not be 100% harm-free as some people may develop allergies or have other side effects. Nevertheless, while it is true that sometimes vaccines can lead to serious adverse effects, this is extremely rare. The minimal risk of an adverse effect should be weighed against the significant protective benefits that vaccines provide.[6] Millions of lives have been saved during past decades thanks to vaccines. A recent analysis has shown that routine childhood immunization among members of the 2009 US birth cohort is expected to prevent about 42,000 early deaths and 20 million cases of disease.[7] Overall, it is well-documented that licensed vaccines are effective and safe for the vast majority of people who use them.

Nevertheless, it is interesting to look at some uncertainties related to the safety of vaccines and at how they have been distorted and exaggerated.

Uncertainties Inherent in Vaccination

As already mentioned, the uncertainties inherent in vaccination have to do with the fact that the same vaccine will not have the same effect on everyone who receives it. Some of the vaccinated individuals, always a minority, will develop a less effective immune response than others, and others, again a minority, may exhibit some side effects that in rare cases could be severe. Due to these rare instances, in recent years an active anti-vaccination movement has emerged in the United States and elsewhere, causing unjustifiable concerns about vaccines. According to Paul Offit, chief of the Division of Infectious Diseases at the Children's Hospital of Philadelphia and professor of pediatrics at the University of Pennsylvania School of Medicine, the birth of the modern anti-vaccine movement in the United States took place on April 19, 1982, when a documentary called *DPT: Vaccine Roulette* was aired ("DPT" stands for "diphtheria, pertussis, tetanus").[8] The main point made in that documentary was that the pertussis (also known as the "whooping cough") vaccine was causing permanent brain damage, epilepsy, and mental retardation in children. Even though there was no solid data to support this claim, which did not make sense anyway in terms of the physiology of the brain, the effect was tremendous in terms of how many parents became concerned and decided to refrain from vaccinating their children.[9]

Interestingly, as Offit noted, no such reaction existed in past cases of real vaccine failure. For instance, in 1954, a polio vaccine that had been developed by Jonas Salk and his colleagues was tested on more than 1.3 million children. Unfortunately, one of the pharmaceutical companies that produced the vaccines, Cutter

Laboratories at Berkeley, California, had failed to completely in-activate the virus. A plausible explanation for what happened was that the virus mixtures remained too long in storage, resulting in sedimentation that led virus particles to clump together. Thus, some of these particles were shielded from the formaldehyde that was added to destroy them, and they remained active, causing the disease in some of the children who received the vaccine.[10] As a result, 120,000 children were injected with failed vaccines; among those, 70,000 suffered a mild form of the disease, 200 were per-manently paralyzed, and 10 died.[11] This caused a lot of concern and distrust in the polio vaccine developed by Salk, but not to the extent that *Vaccine Roulette* influenced the public. As a result of parental decisions to not vaccinate, pertussis cases in the United States increased from 1,895 in 1982 to 4,195 in 1986, generally being higher after 1982 than before.[12]

A remarkable case from a scientific point of view was the re-action caused by the 1998 publication in the prestigious medical journal *The Lancet* of an article that linked measles, mumps, and rubella (MMR) vaccine to autism. Andrew Wakefield, a physician working in the United Kingdom at the time, was the first author of an article noting that "In eight children, the onset of behav-ioural problems had been linked, either by the parents or by the child's physician, with measles, mumps, and rubella vaccination."[13] The behavioral diagnosis was autism. However, the authors were quite ambivalent in their conclusions about the connection be-tween the vaccine and autism. On their last page, they repeatedly referred to the connection between the vaccine and autism, while also exhibiting doubt about it. Thus, whereas they wrote that "We did not prove an association between measles, mumps, and rubella vaccine and the syndrome described. Virological studies are underway that may help to resolve this issue," they immediately added that "If there is a causal link between measles, mumps, and rubella vaccine and this syndrome, a rising incidence might be anticipated after the introduction of this vaccine in the UK in 1988." They concluded by writing "We have identified a chronic enterocolitis in children that may be related to neuropsychiatric dysfunction. In most cases, onset of symptoms was after measles, mumps, and rubella immunisation. Further investigations are needed to examine this syndrome and its possible relation to this vaccine."[14] All in all, whereas the authors expressed some doubt about the connection between MMR vaccination and autism and called for further research, they clearly pointed to such a connection.

The scientific reactions to this study were immediate. Already in the same February 1998 volume of *The Lancet* there was a critical commentary by Robert Chen and Frank DeStefano from the Center for Disease Control and Prevention in Atlanta, Georgia. They made several important points and raised concerns about the validity of the conclusions of Wakefield and his colleagues. The first point was that since the mid-1960s hundreds of millions of people worldwide have received vaccines containing measles without exhibiting either chronic bowel or behavioral problems. The second point was that the Wakefield study was based on a specific group rather than a randomly selected sample from the general population. A false attribution is possible in this case because autism becomes evident at about the age when children are vaccinated. Therefore, it is likely that, in some cases, the symptoms will appear after vaccination, thus exaggerating the importance of the association.[15] Last but not least, in the very same volume of *The*

Lancet there was a report that "measles virus genome is not present in gut mucosal biopsies from patients with Crohn's disease or ulcerative colitis," citing various studies by Wakefield and colleagues who had reported such a connection.[16]

Several other researchers expressed concerns in the subsequent volume of *The Lancet*.[17] Wakefield replied to his critics in that same volume by noting that parents were right to have associated the onset of behavioral symptoms in their children with the MMR vaccine and that they could not be ignored just because this challenged the public health dogma concerning MMR vaccine safety. He concluded his reply by noting that: "Assumptions of vaccine safety, based upon inadequate safety trials and dogma contribute largely to confusion and public loss of confidence in vaccination. Public-health officials would do well to get their own house in order before attacking the position of either clinical researchers or *The Lancet* for what we perceive as our respective duties."[18]

But in 2004 Brian Deer, a journalist who worked for *The Sunday Times,* published an article revealing that Wakefield had different kinds of conflict of interest related to the 1998 *Lancet* study. The title of the article was "Revealed: MMR Research Scandal," and it exposed Wakefield in several ways.[19] Two years before his 1998 *Lancet* article was published, Wakefield had been hired to attack the MMR vaccine by Richard Barr, a lawyer who was hoping to raise a speculative class action against the drug companies that produced the vaccine. Wakefield had received an initial amount of £55,000, which he had not disclosed while his article was under review at *Lancet*. Furthermore, in June 1997, Wakefield had filed a patent on his own measles vaccine while he was continuously referring to such a vaccine as being safer than the triple MMR vaccine.[20] Because of these and other revelations about Wakefield, the General Medical Council decided in 2010 that "Wakefield's name should be erased from the medical register."[21] Eventually, the Wakefield et al. 1998 *Lancet* article was retracted.[22]

Despite these denunciations of Wakefield's purported findings there was an enormous public response. Medical doctor and science writer Ben Goldacre described it as "The media's MMR hoax," which "is the prototypical health scare, by which all others must be judged and understood." According to Goldacre, it was not Wakefield who was responsible for this but rather "the hundreds of journalists, columnists, editors and executives who drove this story cynically, irrationally, and willfully onto the front pages for nine solid years."[23] Media coverage peaked in 2002, with more than 1,000 stories devoted to MMR in national newspapers.[24] But, as Goldacre explained, these stories were not about experimental evidence but rather about didactic statements from authority figures on either side of the debate, giving the impression that the answers were to be given by some experts rather than by a careful analysis of the available evidence. Worse than that, the media circulated stories about findings that were never published in peer-reviewed academic journals, as any credible research should have been. While other important findings that discredited Wakefield's claims, such as that no trace of the measles virus was found in children with the same features as those in the 1998 Wakefield study, were not circulated.[25] The media also failed to publicize the fact that several studies since Wakefield's had found no association between the MMR vaccine and autism.[26]

The story of the MMR vaccine reveals many uncertainties that are inherent in vaccination. Like all medical interventions, things may go wrong in individual cases. But one needs to look at the big picture. Vaccines can have side effects, and once a vaccine is introduced it takes years, perhaps decades, to obtain evidence about whether it causes significant problems. The more people who receive a vaccine, the higher the probability that there will be some problems. In light of this, correctly communicating the issue of vaccine safety to the public is very important so that people can weigh the risks of side effects against the probability of being infected with a deadly microbe. For instance, the vaccines RotaTeq (Merck) and Rotarix (Glaxo-SmithKline) have been extensively used to prevent rotavirus, the most common cause of severe childhood diarrhea worldwide, which has been deadly in some cases. The result has been a significant decrease in hospitalizations and emergency room visits for rotavirus (e.g., by more than 80% among immunized children in the United States), as well as in deaths from diarrhea (e.g., by more than 40% in Mexico). Yet in some cases a small but significant increase in the risk of intussusception (when part of the intestine folds into the section immediately ahead of it) has also been detected. In the United States, there is evidence that intussusception can occur as a result of vaccination with either of these two vaccines in approximately 1–5 cases per 100,000 infants. But when this risk is weighed against the hospitalizations, emergency room visits, and deaths from diarrhea, it is considered acceptable. For the 4.5 million babies born each year, vaccination against the rotavirus is estimated to prevent approximately 53,000 hospitalizations and 170,000 emergency room visits for diarrhea at the expense of causing 45–213 cases of intussusception.[27] The risk of this adverse effect makes it essential to pay attention to how things develop so that appropriate actions are taken in time, if necessary. That being said, the risk involved with this sort of vaccination does not outweigh the significant benefits that it provides. Uncertainty is inevitable, but we can make sound decisions by looking at the evidence concerning the risks and benefits of vaccines.

How Should We Deal with the Uncertainties Inherent in Vaccination?

A reasonable answer to the question of how to deal with uncertainties in vaccination is simply to trust the consensus among experts that licensed vaccines are generally safe despite the rare occurrences of side effects. Of course, one might claim that the views of experts are guided by or serve the interests of the companies producing and selling those vaccines. In the film *VAXXED*, produced in 2016, Andrew Wakefield and his companions suggested that the Center for Disease Control and Prevention actually misled the public about the safety of vaccines by covering up the findings on the link between the MMR vaccine and autism.[28] But how likely it is that all or most scientists in a field are participating in a conspiracy to promote the use of unsafe vaccines and that only a few (like Wakefield) stand against that? Beyond this, one can learn more by discussing the details with physicians or by reading and understanding the available evidence. This is best discussed in meta-analyses, as explained in Chapter 5. There are such studies about the use of new adjuvants,[29] as well as about individual vaccines such as the meningococcal B vaccine for which uncertainties remain but which seems to have a safe profile.[30]

Needless to say that a meta-analysis has shown no link between vaccination and autism.[31] Uncertainties related to vaccination exist, but for the licensed vaccines the risks for our children are fairly limited and typically outweighed by the benefits. As Paul Offit put it in the epigraph to this chapter, the choice not to be vaccinated is not risk-free. And the risk of developing a dangerous disease due to not being vaccinated is a lot higher than the risk of having health problems because of vaccination.

8 | Uncertainties in Human Evolution

> We presently know too little, and are always likely to know too little, to be sure where, when and how modern humans originated.
>
> —*Bernard Wood*[1]

What Do We Know About Human Evolution?

Let us begin with the fundamentals. Human evolution is a fact. We have evolved from an ancestral species who was our common ancestor with the apes—who are in turn our closest relatives. The skeletal and behavioral similarities between humans and apes are easy to see. DNA analyses have also confirmed our relatedness with all apes, the chimpanzees being our closest relatives—in fact, they are genetically closer to us than they are to gorillas. Different estimates are available about when the evolutionary divergence between chimpanzees and humans took place, putting it at around 5 million years ago, whereas the common ancestor of humans and chimpanzees separated from the gorillas about 7 million years ago. This entails that for the last 5 million years our species took its own evolutionary path, which resulted in the significant differences that we observe between us and our closest relatives: 46 instead of 48 chromosomes, bipedalism, larger brains, and cultural learning. Nevertheless, an enormous set of similarities indicates our closeness.[2] Does this mean that we are apes? No, it does not. Gorillas, chimpanzees, and orangutans are apes. We have an ape ancestry, but we are not apes ourselves. We have so many significant and distinct features, which evolved since our divergence from apes, that we have a distinct identity. We have an ape ancestry, but we are better described as "ex-apes," in the same sense that we have a fish ancestry, but we would never describe ourselves as fish.[3]

What has happened since our divergence from the apes? Even though we are currently the only human species living on Earth, we know that several other human-like species have existed in the past, such as *Homo ergaster, H. erectus, H. heidelbergensis, H. neanderthalensis, H. habilis, H. rudolfensis,* and *H. floresiensis.* Scientists have found and studied fossils that have provided important information about the structure and habits of these groups. Overall, it seems that modern humans first appeared in Africa about 200,000 years ago (perhaps even 300,000 years ago)[4] and then spread across the world. However, despite the large amount of evidence we have about the origin of modern humans, there is still a lot that we do not know about the details. The current picture as it emerges from fossils is rather fragmented and, at the same time, indicates the considerable diversity in our lineage during the past 500,000 years (see Figure 8.1). Even though we have evidence of approximately when those other human or human-like species lived on Earth, we do not know the exact relations among them.

What kind of evidence do scientists use to study human evolution? Generally speaking, there are two kinds of evidence: fossils and DNA. Fossils are the remains of the hard parts (usually skeletons) of organisms that have been preserved and have

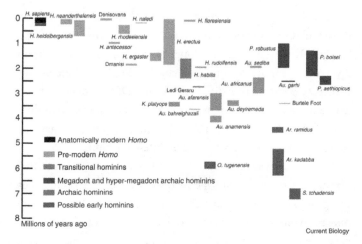

FIGURE 8.1 The currently available evidence for the origin of modern humans is fragmented and filled with uncertainties about the details of human evolution.

Reprinted from Wood, Bernard, "Evolution: Origin(s) of Modern Humans," *Current Biology 27*, no. 15 (2017): R767–R769.

maintained their original shapes. However, many organisms with skeletons have not been fossilized simply because the necessary conditions were not there, and many more organisms—such as most invertebrates—do not have skeletons that can be fossilized. Therefore, fossils are not always available. Yet, when they are, they can provide very important information. For instance, fossils can provide information about the era in which the fossilized organisms lived because it is possible to date the rocks in which the fossils are found by measuring the amount of particular elements constituting the rocks. Other important evidence comes from DNA, in which changes called *mutations* can occur. When these mutations do not adversely affect survival, they can accumulate in DNA. As a result, the longer it has been since two groups diverged, the more accumulated changes in DNA there will be. Hence, two groups that differ less in their DNA sequence should have a more recent common ancestor than two groups that differ more. Scientists can look at these differences and thus estimate the times of evolutionary divergence of different groups.

Our human ancestors coexisted with the Neanderthals (*H. neanderthalensis*) and the Denisovans. Let us consider these two groups in order to make clear the kinds of uncertainties about human evolution that exist. From fossils of Neanderthals that have been found in various places in Europe, Asia, and the Near East, scientists have concluded that some of their distinctive characteristics were a forward-jutting face with a big nose, thick and protruding brow ridges, a projected occipital bone at the back of the skull, and a relatively large cranial volume that is quite similar to our own. Quite interestingly, researchers have also been able to reproduce the genome sequence of the Neanderthals, so there is quite a lot that we know about them. In contrast, we do not know much about the Denisovans because we have only found a few bones. Nevertheless, it has been possible in this case, too, for scientists to study parts of their DNA.

The first Neanderthal fossil bones were found in 1856, in a cave of the Neander Valley in Germany. Researchers were able to obtain and analyze a region of mitochondrial DNA (mtDNA), a small DNA molecule which is found within mitochondria (the organelles that produce energy in cells). This is DNA of maternal origin because only the mitochondria of the ovum and not those of the sperm are included in the embryo emerging after fertilization. Thus, mtDNA is only passed on from a mother to her children. As a result, it can be used to trace one's ancestry to one's mother, one's maternal grandmother, one's maternal great-grandmother, and so on. The analysis of the Neanderthal mtDNA has shown that it was very different from human mtDNA.[5] Further excavations brought to light more Neanderthal bones, which allowed more DNA to be extracted. The analysis showed that this DNA came from an individual maternally unrelated to the first one. The dating of these specimens showed that these individuals lived about 40,000 years ago.[6] These findings provided the basis for a draft sequence of the Neanderthal genome, which was compared to that of humans from various parts of the world. It was concluded that the Neanderthal genome was on average more similar to the genome of present-day humans from Eurasia than it was to those from Africa, and that between 1% and 4% of the genomes of humans in Eurasia were derived from Neanderthals.[7]

There is a lot less data available for the Denisovans. From a finger bone of a hominin found in Denisova Cave in the Altai Mountains in Russia, it was possible to obtain and analyze mtDNA. It was concluded that the Denisovan individual had a common ancestor with modern humans and the Neanderthals, but the Denisovan mtDNA had almost twice as many differences from the mtDNA of present-day humans when compared to the Neanderthal mtDNA. It was also concluded that the Denisovan individual lived between 30,000 and 50,000 years ago, and so coexisted with both present-day humans and the Neanderthals.[8] Further DNA analysis supported the conclusion that the Denisovan individual and the Neanderthals were descended from a common ancestral population that was already separated from the ancestors of present-day humans, and therefore that the Denisovans and the Neanderthals were closer to each other than either of them was to humans.[9] A more recent study has also provided evidence that the Neanderthals, the Denisovans, early modern humans, and a fourth group of an unknown archaic hominin interacted and had offspring on various occasions. It seems that there was transfer of genes (1) from the Neanderthals to the ancestors of many present-day human groups in Asia, (2) from the Neanderthals to the Denisovans, and (3) from an unknown archaic hominin to the Denisovans (Figure 8.2). The available evidence supports

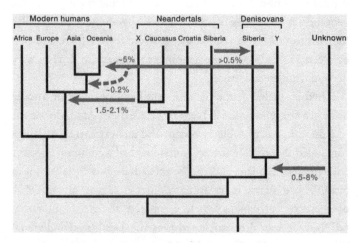

FIGURE 8.2 Possible gene-flow events between the genomes of archaic and present-day humans.

Reprinted from Pääbo, Svante, "The Human Condition: A Molecular Approach," *Cell* (2014): 216–226.

the conclusion that several other humans coexisted and interacted with our ancestors about 30,000–40,000 years ago.[10]

What is one supposed to make of all this? Figure 8.1 is based on fossil data whereas Figure 8.2 is based on DNA data. In both cases, we have fragmented evidence, but we can make particular inferences about the relations between the various groups. DNA data certainly give a clearer picture than fossil data because direct comparisons are more feasible. This is why we can infer which groups are closer to each other in Figure 8.2. In contrast, this is not easy to do with the data in Figure 8.1—this is why the various groups are not connected with lines (even though the shading and placement certainly imply what their relation could be). In Figure 8.1, one could try to draw lines connecting the boxes in various ways. But we would not really know whether the connections drawn were accurate because the available evidence does not support one arrangement over another. This is why there are no connecting lines in Figure 8.1, as one would see in a phylogenetic tree. There are many uncertainties about the exact connections between these human groups, but scientists are confident about the fact that they are genealogically related. The connections in Figure 8.2 are less uncertain. But new evidence might change the picture, especially since the DNA data for the Denisovans has come from just a few bones.

Uncertainties Inherent in the Study of Human Evolution

Uncertainties such as those just described are normal and inherent in the study of human evolution in particular and evolution more broadly. What caused the separation of the human lineage from the others? Under what conditions did the different groups evolve? To answer these questions, we need to have observed the evolutionary processes, but this is impossible. Instead, what scientists can do is search for traces of the effects of such factors in what we see today: fossil bones and the DNA that is found therein. This may sound insufficient, but it is not. Whereas this kind of evidence cannot answer all questions and leaves uncertainties about the exact relations of the various groups, it can nevertheless provide genuine understanding of human evolution. The traces found support inferences about what could have happened and what could not have happened to the extent that we can understand human evolution without having observed it directly as it was occurring. But, of course, we cannot be certain—or even aware—of all the details.

Let us use a simple analogy to illustrate this. Imagine that you return home and find a broken window with pieces of glass all over the floor. Obviously, your window was broken, but since you were not home when it happened, you do not know exactly what happened. Perhaps you left the window open and the wind made it slam and break? Probably not; you always check that the doors and windows are closed before you leave. Perhaps someone threw an object at the window and broke it? Nobody was inside the house, so the object must have come from outside. You then look outside and see some kids playing with a ball. Aha! It could have been them. Perhaps they threw their ball at your window and broke it. Then you wonder: What kind of ball could that be? What is its size? What is it made of? And with what speed did it hit the window? A baseball thrown at high speed at the window could break it. But a ping-pong ball could not. Eventually, you start looking inside your house for a ball. If you do not find one, you may never know what happened. But if you find one, you can infer that those kids playing outside threw it. And if you find one of their names on it, or if their parents confirm that it is indeed their child's ball, then you will know how your window broke. Of course, it might be that not all of the necessary information is available. For instance, if you find a ball but there is no name on it, or if nobody confirms whose it is, you may not know who is responsible for breaking your window. Perhaps some other kids, not the ones you saw, were playing there earlier. Perhaps it was someone else. Nevertheless, the important point is that it is possible to sufficiently explain how your window was broken even if you were not there to observe the event and various uncertainties remain. It must be noted that in cases like this you do not need all the details (e.g., every single piece of glass, or the weight of the ball, or what exactly the ball is made of, etc.) in order to explain why the window was broken. The same principles apply to the study of human evolution. We may not have every piece of evidence, but with sufficient evidence we can figure out what happened.[11]

The study of human evolution and of biological evolution more broadly is filled with uncertainties precisely because the necessary evidence is not always available. Generally speaking, there are three different kinds of uncertainties that scientists confront when they try to reconstruct evolutionary relationships. The first is that extracting DNA from fossils and finding DNA sequences as discussed earlier is not simple or straightforward. DNA may not exist at all in the fossils found, and when it does it may be so degraded that only small DNA fragments can be obtained. Thus, scientists can only obtain fragmented DNA sequences, leaving much of the DNA sequence uncertain. Another problem is that the DNA found in fossils may not come from the organism itself but from other sources such as bacteria, animals, or the humans who found the fossils. A third problem is that, at least for the time being, only a few ancient genomes are available for study. These can in no way be considered representative of the respective populations, and so we cannot really compare the actual DNA variation between contemporary humans with the DNA of extinct humans. Finally, even if we identify that Neanderthals and Denisovans had DNA sequences similar to those of modern humans, we cannot really know if those sequences had the same function or any function at all. Therefore, the inferences that scientists can make from the study of ancient genomes are limited because of these uncertainties.[12]

A final problem is that the relations represented in Figures 8.1 and 8.2 are at the level, or close to the level, of species. In this case, we cannot know which of the groups are distinct species and which ones are populations of the same species, the members of which interbred. Therefore, it is not possible to clearly distinguish between them. As a result, we do not know if the Denisovans are actually a distinct species. Similarly for Neanderthals, experts disagree about whether they are indeed a different species from our own (*H. neanderthalensis*, whereas we are *H. sapiens*) or actually a subspecies of our own species (*H. sapiens neanderthalensis*; in this case, we would be *H. sapiens sapiens*). To this we should also add the consideration that the classification of organisms into species is relatively arbitrary rather than something inherent in nature.[13]

Even when the available evidence is complete, there still exist uncertainties. For instance, scientists have known since the early 1980s that the chromosomes of humans, chimpanzees, gorillas, and orangutans are quite similar. An important difference is that humans have 23 pairs of chromosomes whereas chimpanzees, gorillas, and orangutans have 24. When all these chromosomes were colored with a specialized technique that produces bands of different shades and then compared to one another, a similar pattern of chromosomal bands was found for these species. This entailed that for each chromosome of each of these species there was a corresponding chromosome in the other species that exhibited the same pattern of bands—this is essentially like comparing barcodes and finding them to be nearly identical. The conclusion drawn was not only that these species are closely related, but also what exactly this relation was. From the comparison of the patterns, scientists inferred that chimpanzees and humans are the closest relatives, sharing a common ancestor with gorillas, and all three species in turn share a common ancestor with orangutans. The most striking finding of all was that two individual chromosomes that exist in chimpanzees, gorillas, and orangutans were very similar to the two arms of human chromosome 2. This suggested that a fusion of two chromosomes that existed in the common ancestor resulted in a single, larger chromosome in humans.[14] However, as sure as we are about this event, we cannot know exactly when or why it took place. Similarly, we cannot know when or why our bipedal posture and our larger brains evolved. Uncertainties remain.

As Bernard Wood, a paleoanthropologist at George Washington University, correctly put it in the epigraph to this chapter, there will always be details that we will not know, and thus there is always uncertainty. Epistemic access to the past in the case of human evolution is always indirect and limited. We simply do not always have sufficient and complete evidence, either because it does not exist or because we have not yet found it. And we cannot travel back in time and observe what happened. However, evolutionary processes have left traces that we can examine today and draw conclusions from about what happened. Conclusions come only after the study of the available evidence, and interpretations of the evidence may vary. Thus, experts debate the details among themselves, and this is healthy for science. Uncertainty about the details is what motivates further research and eventually makes science advance. But there is no disagreement among experts over the broad picture of human evolution.

How Should We Deal with the Uncertainties in the Study of Human Evolution?

The uncertainties surrounding human evolution are used by anti-evolutionists as evidence that natural processes cannot account for the emergence of humans, and, therefore, they conclude that other factors have to be responsible. The fact that evolutionary scientists do not (and perhaps cannot) know all the details about human evolution—for instance, exactly how the species in Figure 8.1 are related to one another—is often considered by anti-evolutionists as a deficit of science. Therefore, the only option available, according to anti-evolutionists, is to look to God:

> The only true account of our origins is that provided by God in the Bible. Therefore, the only worldview that can reliably

guide scientists to conclusions about our origin and nature that are *actually true* is one that *does not violate biblical history*—the yardstick by which to assess ideas relevant to the unobservable past. If only all scientists wishing to explain our origins would allow their vision to include an understanding that the physical universe was brought into being by a Creator God, a God who has left us an eyewitness account of our origins and the early history of the earth in Genesis, a history that is consistent with the observable facts of science.[15]

Human evolution is a delicate topic, and its relation to religion, particularly Christianity, is complex. As we have noted, some Christians (and other religious people) are anti-evolutionists, but many others are not. For example, Christian philosopher Richard Swinburne has written that "although there is much uncertainty about the exact stages and mechanisms involved, the fact of evolution is evident."[16] Swinburne is not alone in thinking that evolution is consistent with Christianity. There are several Christians and non-Christians who argue that there is no conflict whatsoever between human evolution and Christian doctrine.[17] Elsewhere we have written about the various emotional and conceptual obstacles that make the idea of human evolution difficult for people to accept.[18] People may think of the idea that we have evolved from animal ancestors as a nihilistic idea, one that deprives their life of meaning. Or, as we have said, they may think that evolution is inconsistent with Christian doctrine. Both worries are misplaced though. For many people, evolution is neither a nihilistic idea, nor is it inconsistent with Christian doctrine. The main implication of the fact of human evolution that we see is simply that we humans should not be arrogant. Human evolution indicates that, despite our many unique features, we are one species among so many others: a short and recent branch in the evolutionary tree or bush of life. Despite the uncertainties in our understanding of human evolution, scientists have no doubt about this fact of life. Therefore, we can appreciate it and let science figure out in detail how we came to be as we are. As in all science, uncertainty in the details of human evolution results in more research that in turn will produce better understanding of human origins.

9 | Uncertainties in Genetic Testing

How can we encourage rapid innovation while ensuring patient safety? By law, the FDA must evaluate diagnostic tests for both analytic validity and clinical validity. For a gene-based test, this assessment comes down to two questions: Does the test accurately read out a targeted set of DNA bases in the human genome? Does the targeted set of DNA bases provide meaningful clinical information?

—Eric Lander[1]

What Is Genetic Testing About?

Eric Lander is an important figure in contemporary genomics research, being the president and founding director of the Broad Institute of MIT and Harvard and having also been in the past a principal leader of the Human Genome Project. In the epigraph Lander raised two very important questions: Do genetic tests measure what they are intended to measure (this is called *analytical validity*) and, assuming they do, is the existence of a particular DNA variant indicative of whether one will develop a disease or not (this is called *clinical validity*)? In that same article, Lander estimated that there exist "3600 genes for rare mendelian disorders, 4000 genetic loci related to common diseases, and several hundred genes that drive cancer."[2] This means that it is likely

121

that each one of us carries several variants associated with diseases. But for how many of these variants should one be tested? Can we be certain that when the test indicates the presence of a particular variant it is really there? And can we be certain that if we have the variant we will develop the disease or that, if we do not have it, we will not develop the disease? A related concern is this: Assuming that the variant is indeed there and that it does indeed cause the disease, is there any medical action that we can take (this is called *clinical utility*)?

These questions are not easy to answer. In this chapter we explain the problems in answering questions about the analytical validity, clinical validity, and clinical utility of genetic tests. Such problems are mostly due to uncertainties inherent in the respective scientific procedures. Before looking at these problems, it is useful to briefly describe the relevant science. In the past, researchers studied people with and without a particular disease in order to see whether particular versions of genes (these are called *alleles*) were more common in one group or the other. If an allele was found to be more common in the "disease" group, it was taken to be somehow involved in the development of the disease. In contrast, if an allele was found to be more common in the "non-disease"/control group, it was taken to be somehow involved in the prevention of the disease. But association does not entail causation; we cannot know whether an allele is involved in the development of a disease or if it prevents the disease until the underlying biological mechanism is understood in detail.

In recent years, this approach to studying genetic influences on diseases has been expanded to include thousands of participants and to search for numerous DNA variants across the whole genome of each participant. These studies are called *genome wide association studies* (GWAS), and they look for differences in the frequencies of variants between different groups. If such a difference is found to be significant for a variant, one may assume that the variant somehow affects the disease.[3] Now the problem is that GWAS have revealed associations among numerous different variants and particular diseases. For instance, GWAS for nearly all common cancers have been performed, leading to the identification of more than 450 variants associated with increased risks for developing cancer. This research has supported the conclusion that common genetic variations contribute substantially to the risk of developing many common cancers, as well as reveal novel pathways important in carcinogenesis. However, with the exception of breast and prostate cancers, the currently identified variants explain only a small proportion of the familial risk of many cancers.[4] Put simply, most of the genetic variants related to cancer identified so far make only a very minor contribution to cancer development.

How do scientists search DNA for these variants? How is it possible to "read" DNA? Broadly speaking, there currently exist two ways. In the procedure known as *whole genome sequencing*, it is possible to determine the sequences of short DNA fragments into which a person's DNA has previously been cut. These are then compared to one another, and it is possible to estimate their order based on their overlapping (common) parts. The final product is an ordered combination of these parts that corresponds to the sequence of the whole genome, which is nevertheless incomplete because some parts of DNA are hard to sequence. To minimize errors, each variant is usually sequenced about 30 times (and a lot more than that when the aim is to identify a disease-associated variant). Another approach is called *exome sequencing*. In this case the aim is to find the sequence of the exons of genes, which are the parts of genes from which a functional molecule such as a protein is produced. This is less difficult and less costly than whole genome sequencing, but it may miss important information about variants that are found outside protein-coding genes.[5] Regardless, the end product of both methods is a DNA sequence. But, while we can certainly read a DNA sequence, making sense of it is a very different story.

Based on technologies like these and the findings of GWAS studies, various companies have developed genetic tests that aim to look for the presence or absence of particular DNA variants and estimate one's probability of developing a disease. One well-known company of this kind is *23andme*. In August 2013, *23andMe* launched a national television commercial, which suggested that people could learn a lot about their health and risk for diseases for $99.[6] But in November of the same year, the Food and Drug Administration (FDA) ordered *23andMe* to stop marketing the respective kit until it received authorization. The reason for this decision was that *23andMe* had failed to respond to questions from the FDA about the analytical validity and the clinical validity of the test.[7] However, as of April 2017, *23andme* was allowed to sell its Personal Genome Service Genetic Health Risk (GHR) tests for 10 diseases or conditions.[8] Currently, the home page—accessible in the United States—states: "What can your DNA say about your health? Learn more about your health, traits and ancestry, with a package of 75+ reports that only the 23andMe service offers." This service offers a number of reports, including five reports for "Genetic Health Risks" that meet FDA requirements.[9]

Another company, *Futura Genetics,* states on its introductory page: "A healthier future starts now. Discover and reduce your likelihood of developing 28 common conditions with Futura Genetics DNA test."[10] Yet another company, *Pathway Genomics*, goes even further by stating that "DNA holds the blueprint to how your body responds to the world around it. Because you are unique, the traditional 'one-size-fits-all' approach simply doesn't work. Together, we'll uncover your personal path to a healthier lifestyle."[11] Referring to DNA as a sort of blueprint, as these companies do, is making a very strong claim because it entails that there is some information or message already prescribed in DNA. The implicit message from these and other statements found on the websites of companies selling genetics tests is that there is knowledge hidden in your DNA, and you had better become aware of it. But how easy is it to get this knowledge?

Uncertainties Inherent in Genetic Testing

There are many different kinds of uncertainties in the technologies used in genetic testing. As a result, the fact that an association exists between a particular DNA variant and a disease does not entail that a person who has the variant will definitely develop the disease. First of all, given the methods used, when thousands of DNA variants are tested one might find an association between a variant and a disease just by chance—that is, the apparent association may not be a sign of a real biological connection at all. The parameter of significance most often used in science is the *p-value*. A p-value indicates the probability that a particular result is due to chance rather than a genuine association. For instance, a p-value of 0.05 means that when it is taken to be true that there are associations between particular variants and a disease, it is still 5% (0.05) probable that an association found exists just by chance and that it does not track any real link between the DNA variant and the disease at all. For instance, if a study concludes that for 500,000 variants there is a statistically significant association between those and the disease, then, given the typical p-value, we should expect that about 25,000 of the variants found to be statistically significant are not really tracking genuine links between DNA variants and disease. In such a case claims of statistical significance could seriously mislead. This is why a p-value of 5×10^{-8} or 0.00000005 is required in GWAS studies to conclude that an association is significant. And still this is insufficient for confidently concluding that uncommon variants (those found in less than 5% of the population) in populations of European ancestry are genuinely associated with disease. Things are even less clear for populations of non-European ancestry who are less well studied.[12]

Let us now assume that the association between a particular, common DNA variant and a disease is accurately established. Are we certain that the results of a genetic test are accurate? Unfortunately, no. Analytical validity is about how accurately a test identifies a particular genetic variant. Ensuring the analytical validity of a genetic test and understanding the accuracy, precision, and limits of the detective capabilities of that test is critical. Analytical validity depends both on the means, such as instruments, reagents, and software, used and on the laboratory processes for using those means. As these may differ from one laboratory to another, it is possible that the analytical validity of the same test varies from laboratory to laboratory. Once a variant is considered to be accurately detected, we then need to consider clinical validity, which is how well a test identifies a person who carries the variant in question. There exist particular properties that indicate clinical validity:

- *Sensitivity*: The proportion of people who have a positive test result among those who carry the variant.
- *Specificity*: The proportion of people who have a negative test result among those who do not carry the variant.
- *Positive predictive value*: The proportion of people who carry the variant among those with a positive test result.
- *Negative predictive value*: The proportion of people who do not carry the variant among those with a negative test result.[13]

This is complicated! Let us break things down a bit more. There are two possible pitfalls. On the one hand, a genetic test may indicate that one carries a disease-related variant when one actually does not. This is called a *false-positive* result. On the other hand, a genetic test may fail to identify a disease-related variant even though one does actually carry it. This is a *false-negative* result. These two kinds of errors are reflected by two important features of a test: its sensitivity and its specificity. The sensitivity of a test indicates the probability that a test yields a true-positive result. If the sensitivity of a test is 99%, this means that 99 out of 100 tests performed will accurately indicate that people have the DNA variant that the test is intended to identify. However, 1 of these 100 tests will show that the variant is there when it is not, a false-positive result. The specificity of a test is the probability that a negative test result is truly negative. If the specificity of a test is 99%, this means that 99 out of 100 tests performed will accurately indicate that people do not have the respective variant. However, 1 of these 100 tests will show that the variant is not there even though it is, a *false-negative* result.[14]

Why does this matter? Let us consider an example. On March 6, 2018, the FDA gave authorization to *23andMe* to market a direct-to-consumer genetic test for three *BRCA1* or *BRCA2* mutations.[15] Let us assume that the sensitivity of this *BRCA1/2* test (true positives) is 99.99% and that the specificity of this test (true negatives) is also 99.99% (which would be really quite high for a test of this sort). Can we be certain about the results of this test? Well, there are two kinds of problems here. The first is that the three mutations in this test are found in 2% of women of Ashkenazi Jewish descent, but only in 0.1% or less of women from other populations. Given the low frequency of these mutations in other populations, there is the danger that women taking this test could get negative results simply because these three mutations were not found. Thus, they might be falsely assured that they have no risk of developing cancer even in spite of a family history of cancer. Second, even by assuming a 99.99% sensitivity and specificity for the test, it is possible for women coming from populations in which these mutations are rare (less than 0.1%) to get false-positive results: it is estimated that 1 in 10 positive results would be false positives in that case, a number that would increase if the sensitivity and the specificity of the test were lower.[16]

There are additional issues with the validity of direct-to-consumer genetic tests. Researchers looked at raw DNA data provided by companies performing these tests, coming from 49 people from whom the companies requested further testing. The researchers found that 40% of DNA variants in several genes reported in the raw data were false positives. Furthermore, eight variants in five genes were designated as "increased risk" in the companies' raw data; however, several clinical laboratories had classified these variants as benign, and publicly available databases mentioned them as common variants. The researchers noted that self-tests are less reliable than what they should be, and their findings should be confirmed in clinical laboratory settings that have the required expertise in the detection and classification of complex variants. The researchers noted that "While the raw data include disclaimers stating that they have not been validated for accuracy and are therefore not intended for medical use, they could easily be misinterpreted or misused by a consumer or medical provider with little to no training on the complexities of genetics."[17]

Now, let's assume that both the association between a DNA variant and a particular disease is valid and a genetic test can accurately indicate that a person carries this variant. Can we now be certain whether the person will develop the disease or not? As we saw in Chapter 2, based on the data provided by the American Cancer Society, women with mutated *BRCA* alleles who will develop breast cancer by the age of 70 make up significantly less than 1% of the general population. What can we make of this? Well, women who carry these alleles should certainly be concerned. But even then, things are not simple and straightforward. As we saw earlier, according to the American Cancer Society, the average probability for a carrier of *BRCA1* mutations to develop breast cancer is estimated to be between 57% and 65%, whereas for *BRCA2* mutation carriers the odds of developing breast cancer are estimated to be between and 45% and 55%.[18] This simply means that among 10 women with the *BRCA1* mutations, about 3–4 will not develop breast cancer, and, among 10 women with the *BRCA2* mutations, about 4–5 will not develop breast cancer. Therefore, finding out whether one has a variant or not, even for these cases, can only provide a probabilistic estimate far short of certainty. This means that in several cases the clinical validity of a genetic test is limited, and none provides absolute certainty.

Worse than this, *BRCA1* and *BRCA2* are not the whole story when it comes to breast cancer. In a study that included 174 women with breast cancer and 24 women without, several disease-related variants and variants of unknown significance were found to exist outside the *BRCA1* and *BRCA2* genes. Overall, mutations in *BRCA1* and *BRCA2* genes were found in approximately 28% of the participants, and other disease-related variants were found in approximately 8% of the participants. It is far from simple and straightforward to figure out the best set and number of DNA variants that should be included in a test.[19] On the basis of all this, it should be clear that genetic testing about cancer and other complex diseases can provide only limited information. Having mutations in *BRCA1* and *BRCA2* genes does not guarantee that one will develop breast cancer, and lacking such mutations does not guarantee that one won't. Certainty simply is not to be found here.[20]

Last but not least comes clinical utility: whether a test can provide information about diagnosis, treatment, management, or prevention of a disease that will be helpful to people. Assuming that we do find a DNA variant that is clearly related to a disease. What can we do? A study assessed the psychological, behavioral, and clinical effects of the use of direct-to-consumer genetic tests by looking into the results of using the Navigenics Health Compass. Among those who had used this commercially available test, 2,037 people reported whether or not they experienced changes in symptoms of anxiety, intake of dietary fat, exercise behavior, test-related distress, or the use of health-screening tests approximately 6 months after testing. The researchers concluded that using the genetic test did not result in any measurable short-term changes in psychological health, diet, exercise behavior, or use of other medical tests. In other words, the knowledge that a genetic test is supposed to confer does not necessarily result in a change of attitude that might lead to disease prevention.[21] The reason for this is that if we do not know the underlying mechanisms in detail—simply put, what the DNA variants do—then there is not much to recommend. Hence, even if we were certain that someone has a particular DNA variant and also were certain that this is associated with a disease, we would still be uncertain as to the best steps for the person to take to avoid the disease.

How Should We Deal with the Uncertainties in Genetic Testing?

Despite the uncertainties and the problems just mentioned, genetic testing can provide meaningful information. The International Cancer Genome Consortium is an initiative launched to coordinate large-scale genome studies of tumors in thousands of genomes from 50 different cancer types.[22] The study of the genomic profile of tumors is expected to help find markers in the genome that might allow for the identification of cancer at an early stage of development, when effective treatment is still possible. At the same time, the study of the genomic profile of tumors in relation to the lifestyle habits of the people from which these come might help to identify possible carcinogens and eventually to figure out how to limit human exposure to various harmful substances.[23] This is far from simple and straightforward, but with time it might provide clinically useful information. Unfortunately, we are not there yet. As George J. Annas, William Fairfield Warren Distinguished Professor at Boston University, and the late Sherman Elias, Professor Emeritus in the Department of Obstetrics and Gynecology of the Feinberg School of Medicine at Northwestern University, put it: "For the immediate future, however, we will only be able to probabilistically predict, but not prevent, most diseases."[24] The reason for this is all the uncertainties inherent in genetic testing. The question for us becomes: What are we to make of all this, and how do these uncertainties impact science?

These kinds of uncertainties do not seem to affect people's views in the way the uncertainties surrounding climate change, vaccination, and human evolution do. Whereas in those cases people may refer to the existent uncertainties as reasons to question the science, in the case of genetic testing people seem to do exactly the opposite: ignore the uncertainties and try to acquire personal information using the science. In recent years, many of the companies selling genetic tests have focused on ancestry tests. During 2017, the number of people who had their DNA analyzed exceeded 12 million. The company *Ancestry.com* tested more than 7 million people, followed by *23andMe*, which tested more than 3 million people. Perhaps you will not be surprised to read that these two companies also spent more than any others in advertising during 2016: *Ancestry.com* spent $109 million, followed by *23andMe* which spent $21 million.[25] Here is a case in which the uncertainties inherent in science do not matter to people as much as they should. People are eager to spend money hoping to learn more about themselves, ignoring the uncertainties inherent in the technologies used and in the inferences drawn.

10 | Uncertainties in Forensic Science

> But I believe that blind procedures would also reduce the temptation to falsify DNA data, misrepresent findings in laboratory reports, and ignore evidence of problems in assays. Analysts who do not know whether their tests point to the "right" person will (I believe) be more cautious and rigorous in their interpretations and more honest in acknowledging uncertainty and limitations in their findings.
>
> —*William C. Thompson*[1]

What Is Forensic Science About?

Broadly speaking, forensic science consists of those scientific methods, tools, and approaches used in criminal investigations to obtain evidence that is not readily available. The classic example of such evidence is fingerprints, which have a unique pattern for each one of us. The idea is simple: if there is a database of everyone's fingerprints and fingerprints are found at a crime scene, then comparison of the crime scene fingerprints to those in the database could result in a positive match with a particular person. The person whose fingerprints were found at the crime scene could thus—in principle—be considered a suspect, or so the argument goes. In recent years, the same basic idea has been expanded to comparisons of DNA, which was initially described as

"DNA fingerprinting." More recently, other terms have been used such as "DNA profiling," "DNA typing," and "DNA testing." In this book, we use the term "DNA profiling" to refer to DNA analyses conducted in the context of criminal investigations. In DNA profiling, DNA is extracted from human tissues, such as blood, saliva, hair, or semen, found at a crime scene, and is later analyzed. This results in a DNA profile. Then, insofar as DNA databases are available, it is possible to compare the DNA profile from the crime scene with those in the database. The goal is to match the DNA from the crime scene with a particular person.[2] The use of DNA in criminal investigation has been generally considered as "emblematic of a level of objectivity and certainty," unmatched by any other kind of evidence: for some it is a *truth machine*.[3]

Thousands of people have been convicted of crimes and sent to jail on the basis of DNA evidence. But perhaps more striking are the cases of people who have been exonerated on the basis of DNA evidence. This is the goal of the Innocence Project: "The Innocence Project, founded in 1992 by Peter Neufeld and Barry Scheck at Cardozo School of Law, exonerates the wrongly convicted through DNA testing and reforms the criminal justice system to prevent future injustice." As of our writing, it is estimated that 356 people in the United States have been exonerated through DNA testing, 20 of whom were on death row. On average, the exonerated people had served 14 years in prison.[4] On the other side, there are cases in which DNA leads to a suspect, as in the recent case of the so-called Golden State Killer who is thought to have committed more than 50 rapes and 12 murders across California in the 1970s and 1980s. The identity of this person was a mystery for decades. Recently, criminal investigators submitted DNA that had been collected many years ago from a crime scene linked to the Golden State Killer to one or more companies that have big databases of genetic information. Eventually, the results led investigators to some people who had submitted their own DNA in order to find out more about their ancestry, and the subsequent search led to a relative of theirs, a 72-year-old former police officer.[5] What will happen remains to be seen. Nevertheless, the possibility of using DNA to bring guilty people to justice and exonerate innocent people seems really impressive.

Successes like these may serve as evidence for the importance and usefulness of DNA profiling. However, a myth of infallibility concerning DNA forensic testing has been propagated, perhaps most effectively by popular TV series such as *CSI: Crime Scene Investigation*. The show's lead character, Gil Grissom, makes several statements that reflect a specific view of forensic evidence: "We're crime scene analysts. We're trained to ignore verbal accounts and rely instead on the evidence a scene sets before us.... I tend not to believe people. People lie. The evidence doesn't lie."[6] What is the message here? That we should look at biological evidence, which cannot lie and which is sufficient on its own to reveal the truth beyond any doubt. Indeed, a detailed study of this TV show concluded that the dominant message conveyed is that DNA profiling is common, swift, reliable, and instrumental in solving cases. It was also found that the show presents DNA analyses as quick and easy, typically taking no more than a day to complete, whereas in reality this process may actually take days, weeks, or even months.[7]

In this chapter we focus on the uncertainties inherent in forensic science. Our aim is to show that even if we had vast DNA databases with information about every single person in the world and even if we all agreed that we can use these databases, there would still be unresolved issues. The techniques used in DNA profiling are quite different from the genomic analyses that we described in the previous chapter. An important difference is that, for DNA profiling, it makes no difference where genes are or whether the DNA sequences analyzed relate to some biological trait or not. Rather, what matters is that DNA sequences that vary a lot among individuals are compared. Humans' DNA sequences are very similar, perhaps more than 99.5% alike. Therefore, for DNA profiling, it makes no sense to examine the whole genome of individuals. Instead, what is examined are DNA sites that belong to that 0.5% of differences, which corresponds to approximately 15 million nucleotide bases (represented with the letters A, T, C, and G) of our 3 billion nucleotide base genome. These sequences are usually found outside genes, in sites which are nonfunctional and which can thus accumulate changes and vary significantly among individuals. The differences occurring can either be in the order of nucleotide bases or in the number of times that a certain DNA segment is repeated. What DNA profiling techniques do is compare these sequences and provide the means for distinguishing among different individuals based on their variability.

The initial method used, DNA fingerprinting, was based on the differential cutting of DNA molecules with specific enzymes that can "recognize" and "cut" only very specific DNA sequences. If these sequences differ significantly among different individuals, then the DNA of each one of them will be cut at different sites and produce DNA fragments of different length. Then, under the influence of an electric field, the negatively charged DNA molecules can move toward a positive electrode within a porous gel. Because the shorter fragments move faster inside the porous gel than the longer ones, they can be separated and made visible. Thus, an image resembling a barcode can be produced for each individual.[8] More recently, other methods have been developed. One produces the *electropherogram*, which is the output of a DNA analyzer showing the alleles for each individual and which is based on a standard set of short repeated sequences in human genomes. In this case, alleles are indicated by peaks, and a person can have one peak if he or she has inherited the same allele from each of his or her parents or two peaks that each correspond to one of the alleles inherited from each of the parents. Another method is the *IrisPLEX assay*, which requires only six genes to differentiate among 40 shades of blue or brown eye color.[9] These high-tech methods are impressive and very accurate, but they are not infallible. Important sources of error exist, and several different kinds of uncertainties are inherent both in the methods used and in the interpretation of the available evidence. Misunderstanding of these uncertainties has resulted in wasted time, effort, and money in criminal investigations.

Uncertainties Inherent in Forensic Science

Many uncertainties are inherent in DNA profiling methods. DNA is extracted from tissues through a process that removes proteins, lipids, and all other biological molecules, resulting in pure DNA. This can be a relatively simple process, but it does not entail that the quantity of DNA emerging will be appropriate for analysis. If too much DNA is extracted, then it is possible to obtain unclear results (e.g., off-scale peaks in the electropherogram); if too little DNA is extracted, then some peaks will not appear, thus producing misleading results (e.g., someone carrying two different alleles might appear as though they only have one). Another problem is that the DNA molecules of two different people are sometimes mixed—a cause of seriously misleading results. A third problem is that DNA breaks down over time, depending on the tissue of origin and the conditions of storage. Thus, if DNA is analyzed long after a crime, the results obtained might be quite misleading.[10]

Let us set aside worries about the quality of the DNA sample and assume that the results are completely accurate. Even in that case, there is the issue of interpretation. Let us consider an example of DNA fingerprinting. Figure 10.1 presents what could be

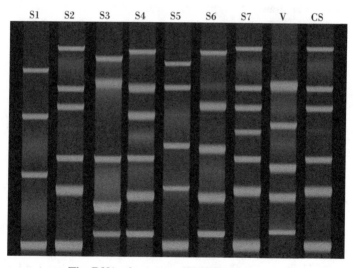

FIGURE 10.1 The DNA of suspects (S1–S7) and the victim (V) compared to that found in the crime scene (CS). Do you think you can find who committed the crime?

perceived as the DNA barcodes of several individuals who might be related to a crime, as well as that of DNA found at the crime scene that does not belong to the victim. Can you infer who committed the crime?

If you answered that the person who did the crime is S2, you are wrong! Not because this could not be the person who did it, but because you cannot legitimately infer this from the DNA evidence. Even though the DNA profile of S2 and CS seem to be almost identical, this does not in any way "prove" that this person is the criminal. It does not even "prove" that it was this person's DNA at the crime scene. DNA alone cannot support this conclusion for reasons that we explain later. A better supported inference that can be drawn from this image is that the DNA found at the crime scene does not come from suspects S1 and S3–S7. Again, this does not mean that these people were not involved in the crime. It only establishes that the DNA found at the crime scene does not belong to them. One would need a lot more, and very different kinds of, evidence than what is provided by this DNA in order to reach a juridical decision.

DNA evidence does not speak for itself; it is all a matter of interpretation. William C. Thompson, Professor Emeritus of Criminology, Law, and Society; Psychology and Social Behavior; and Law at the University of California, Irvine, has noted there are important human errors that are possible:

- *Cross-contamination* of samples, as in the case of the Phantom of Heilbronn in Germany, where the police were looking for years for a person whose DNA had been at several different crime scenes; it turned out that the DNA came from a woman who worked in a factory that produced the cotton swabs used to collect DNA from crime scenes.

- *Mislabeling* of samples, as in the case of the Night Stalker, a person involved in sexual assaults in London, whose arrest was delayed because of an error that excluded him from the investigation and involved another man with the same name.

- *Misinterpretation* of test results, as in the case of a man in Sacramento, California, who was investigated for being involved in a crime because of a bad interpretation made by the analyst in the forensic laboratory.

These uncertainties are due to human errors: someone did not take the necessary precautions for avoiding contamination, confusing the samples, or making a bad interpretation of the findings. Coincidences and manipulations are also possible. The DNA of S2 in Figure 10.1 may have been found at the crime scene because S2 happened to be there independently of the crime or because someone else tried to frame S2.[11] The human factor can thus be responsible for a lot of mistaken interpretations in DNA profiling. But it is not the only culprit. Assuming that no human errors occur, there still exist problems that are due to uncertainties in the methods used themselves.

There are important issues related to the statistics used in forensic analyses. When forensic experts give their reports, they also provide the probability of a genuine match. Where do these probabilities come from? These are usually described as *random match probabilities* (RMP)—the probability that two different people happen to have the same profile. This probability is calculated by multiplying the individual allele frequencies of independent DNA sequences (found on different chromosomes). The greater the number of DNA sequences and the greater the number of different versions of each sequence, the lower the probability of a random match is. However, this probability can be significantly increased if the profile is partial because of DNA degradation, if a suspect and a perpetrator share many alleles because they are relatives, or if a suspect and a perpetrator originate from the same subpopulation.[12]

Let us assume that in a particular case the probability of a random match is 1 in 1,000,000. How likely is it to find someone else with the same DNA profile just by chance? At first glance this seems very unlikely. But it is not. Even if the DNA of S2 in Figure 10.1 seems to be the same as the DNA found at the crime scene, it is possible that it actually is not. The crime scene DNA might come from another person who happens to have the same profile as S2 by coincidence. Let us explain. Assume that an expert analyzed the case presented in Figure 10.1. The report produced would not include any conclusion about whether S2 is guilty or not. Rather, the report would state that the probability that the DNA found at the crime scene comes from S2 is 99.9999%. This means that the probability that this is not the case and that the DNA comes from someone else is 0.0001%, or 1 in 1,000,000. This might make a coincidental match seem quite unlikely, but the context matters. The statement that the probability that a DNA match occurred by chance is 1 in 1,000,000 simply means that 1 in 1,000,000 people is expected to match by chance. If we are talking about a town of 10,000 people, this would be very unlikely. But it would be a lot more likely in a big city of 10,000,000 people, where there could be 9 other people with the same DNA profile.[13] Therefore, the interpretation of DNA profiling is very context-dependent.

There is another problem that might easily be overlooked. In order to perform a genetic test, one needs some cells, usually blood cells or epithelial cells from our saliva. As every cell in one's body has resulted from the subsequent cell divisions of an initial fertilized egg, in theory all cells of that person's body should have the same DNA. But this is not necessarily the case. The reason for this is that copying mistakes can occur during the replication of DNA that takes place before each cell division. These mistakes can result in DNA molecules with slightly different sequences in the cells emerging from each cell division. Mistakes can also occur during the segregation of DNA molecules that takes place during each cell division, resulting in variation in the structure or the number of DNA molecules in each cell. Consequently, since these mistakes can take place at any stage of development, an individual person can end up consisting of a collection of cells that have differences in the DNA that they contain, a phenomenon known as *genome mosaicism*. This is very important because if a test is based on the analysis of DNA from a few cells, it could be the case that the DNA profile corresponds to the sampled cells but not the whole individual.[14] A related phenomenon is *chimerism*. A chimeric human, or human chimera, is a person who has two or more populations of genetically distinct cells that originated from different zygotes. This usually happens during embryo development when the cells of what would be dizygotic twins (i.e., twins who resulted from the independent fertilizations of two different ova by two different sperm cells) eventually fuse to form a single embryo. As a result, this embryo develops to become a human who consists of cells with two different kinds of DNA sequences.[15] These cases may be rare, but they do occur. Thus, even setting aside human errors, absolute certainty about the results of DNA pro-filing is simply not possible.

These are some examples of the uncertainties inherent in DNA profiling: the quantity, the quality, and the origin of the DNA compared, as well as the interpretation of the results and the inferences drawn from them. It is not possible to be certain; DNA evidence does not tell us the whole truth; it does not even speak for itself.

How Should We Deal with the Uncertainties in Forensic Science?

The usual representation of forensic science in popular TV shows such as *CSI* is that some biological material is enough to identify a criminal. You take the sample, you quickly analyze the DNA therein, and—voilà: the suspect's photo appears on the computer screen. No crime remains unsolved? Well, not quite. As we saw earlier there exist different kinds of uncertainties that need to be care-fully considered before a verdict is rendered. It is also important to note that DNA evidence is only one kind of forensic evidence; there are others that must be carefully considered as well. Whereas it is true that witnesses may lie or simply forget, DNA can also mis-lead either because of decay or because of the misleading results an analysis may produce or be interpreted as producing. Indeed, inquiries both in the United States and the United Kingdom have expressed concerns about biases in decisions made on the basis of forensic evidence. Therefore, judges and jurors need to be aware of the inherent uncertainties and try to limit the resulting biases.

The only way to deal with these uncertainties is for them to be explicitly discussed and explained in courts. As William Thompson stated in the epigraph to this chapter, forensic analysts themselves should take special care in how they interpret evidence in order to avoid biases.

Neuroscientist Itiel E. Dror nicely summarized other kinds of biases in a recent editorial in *Science*. He noted that forensic experts have to deal with particular kinds of biases. One very important one is exposure to irrelevant contextual information: since they collaborate with the police and the prosecution, forensic experts may come to acquire information that is not directly relevant to their work, such as a suspect's ethnicity or criminal record. This information may make the expert predisposed toward a conclusion. For instance, if the expert believes that people of a certain nationality in a particular area tend to commit more crimes than others, and a suspect both belongs to that nationality and lives in that area, an expert may be inclined to think that the suspect is extremely likely to be the criminal. Irrelevant information can also cause a "bias cascade" from one component of an investigation to another, or even a "bias snowball," where different components of the investigation influence one another. For instance, believing that a person is of a certain ethnicity might make one pay more attention to a witness mentioning that she saw such a person, ignoring other witnesses who mentioned that they saw other people close to the crime scene. Finally, bias may also arise when experts do not let the evidence drive the investigation. For instance, instead of following the evidence in finding the suspect, they may try to find evidence that a particular suspect did indeed commit the crime. Dror made specific suggestions to address these problems. Biases due to exposure to irrelevant information can be avoided if each expert only gets the information that is required for the respective analysis that they will perform. An additional measure to avoid bias cascade and bias snowball is division of labor between experts, such as having one collect the information and another analyze it. Finally, biases toward a particular suspect could be avoided if experts came to know more about that person only after the forensic evidence has been analyzed. [16] But these and other measures can be taken only when judges, jurors, and police chiefs become aware of the uncertainties inherent in forensic science and the potential for biases.

PART III

Accepting Uncertainty in Science

11 | Uncertainty Is Inherent in Science

> The deepest misunderstanding about science . . . is the idea that science is about certainty. Science is not about certainty. . . . The very expression "scientifically proven" is a contradiction of terms. There's nothing that is scientifically proven.
>
> —*Carlo Rovelli*[1]

Is Science Uncertain?

We have seen in the preceding chapters that there are many uncertainties when it comes to the details concerning various questions related to climate change, vaccination, human evolution, genetic testing, and forensic science. After learning of such uncertainties, one might begin to worry whether this entails that these domains are inferior to other areas of science. One might reasonably think that *good* science ought to be certain and that whatever is filled with uncertainties should not be considered good. Unlike the beliefs we hold in our everyday lives (see Chapter 1), one might think that science ought to provide us with knowledge that is not plagued by uncertainty. Since these domains are faced with many uncertainties, perhaps they are not really scientific?

Unfortunately, as mathematician William Byers has explained, thinking that to be scientific is to be certain is "more characteristic of a mythology of science" rather than how science really is.[2] In fact, scientists recognize that there is always going to be some amount of uncertainty in science regardless of the advancements we make in constructing better and better theories, developing superior technologies and devices, and so on. Recognizing that all science is uncertain is important for at least two reasons. First of all, it is simply the truth about science. We cannot really understand science unless we appreciate its uncertainty. Second, this recognition helps us see that the fact that there are uncertainties about climate change, say, does not mean that we should not trust what climate scientists tell us or that we should not accept that humans are significantly contributing to climate change. In this chapter we take a step back from the particular details that we have been exploring in the preceding chapters to consider general features of *all* science that make it uncertain.

Science Is a Human Endeavor

The mere fact that science is done by human beings introduces uncertainty into the picture. All humans are fallible. Aren't scientists better at reasoning than the average person, though? Perhaps, but they are still humans, and, as such, they are susceptible to the same biases that we all are. For example, cognitive scientists Hugo Mercier and Dan Sperber reported that studies have shown that scientists fall prey to "myside bias" just like everyone else.[3] Recall from Chapter 1 that this is the bias where we find it difficult to appreciate evidence that runs counter to our own opinions. As a result, we tend to interpret the evidence differently depending on our prior convictions. Scientists make this same mistake, and the famous physicist Max Planck once claimed that "A new scientific truth does not triumph by convincing its opponents and making them see the light, but rather because its opponents eventually die, and a new generation grows up that is familiar with it."[4] While things may not be quite as bad as Planck suggested, scientists are humans and are therefore vulnerable to biases and errors. Thus, they face uncertainty in their conclusions just like the rest of us.

But there is more than being biased. Scientists may intentionally discredit and disregard the available evidence in order to reach particular conclusions for political, ideological, or other reasons. Sadly, there are numerous such instances. Historians of science Naomi Oreskes and Erik M. Conway have described how particular scientists misled the public about the connection between tobacco and cancer, ozone depletion, and more. In these cases, it seems that particular scientists were either influenced by monetary gain, their political leanings, or both. What happened in these cases was that a handful of men who were well known and highly respected by virtue of their earlier work in the weapon programs of the Cold War presented themselves as experts on topics for which they really had no expertise. And what they did was question the findings of the science of the time by highlighting various uncertainties. How can we say that tobacco smoking causes lung cancer if most people who smoke do not develop lung cancer? Why do some smokers develop lung cancer while others do not? Thus, the tobacco industry, for instance, hired some well-respected physicists, such as Frederick Seitz, to argue that the science was still unsettled about the issue. As Oreskes and Conway explained: "Industry doubt-mongering worked in part because most of us don't really understand what it means to say something is a cause. . . . Doubt-mongering also works because we think that science is about facts—cold, hard, definite facts. . . . Doubt is crucial to science—in the version we call curiosity or healthy skepticism, it drives science forward—but it also makes science vulnerable to misrepresentation, because it is easy to take uncertainties out of context and create the impression that *everything* is unresolved. This was the tobacco's industry key insight: that you could use *normal* scientific uncertainty to undermine the status of actual scientific knowledge."[5]

But even if scientists were completely unbiased, there are other issues stemming from the fact that the natural world is extremely complex. Philosopher Angela Potochnik has identified four specific ways in which our world is complex: (1) there is a large variety of different phenomena, (2) for any given phenomenon there's an enormous range of factors that influence it, (3) the various influences on a phenomenon might differ in important ways from the influences on similar phenomena, and (4) individual influences can affect phenomena in complex ways.[6] Let us follow Potochnik and consider the complexity of the causes of obesity as an example. Take a look at Figure 11.1. This is an attempt to map the various causes of obesity. Each of the nodes represents identified causal influences on obesity, and the lines between them represent different strengths of interaction. The key point to recognize here is that a single phenomenon such as obesity is exceedingly complex. In fact, the condition is even more complex than the diagram illustrates. As Potochnik has noted, in such diagrams only some causal influences are depicted, and each of the represented causal influences is affected by numerous other causal influences that are

Foresight
Obesity System Map

FIGURE 11.1 Obesity System Atlas.

https://assets.publishing.service.gov.uk/government/uploads/system/uploads/attachment_data/file/296290/obesity-map-full-hi-res.pdf. Accessed May 10, 2018. Public-sector information licensed under the Open Government License v3.0.

not shown. Additionally, even though various strengths of influence are depicted, the complications in terms of how these various influences interact are not illustrated. Other seemingly straightforward phenomena and conditions are similarly very complex.

Therefore, when we realize that science is something that is performed by humans who are cognitively limited in important ways and done in an effort to understand things that are tremendously complex, it should not be a surprise that science is filled with uncertainties.

The facts concerning the complexity of the world and the cognitive limitations of humans help to shed light on another essential feature of science: it is thoroughly collaborative. It is in an effort to overcome their own limitations in the face of great complexity that scientists collaborate quite extensively, but uncertainty is a limitation that is not easily overcome. Scientific research is seldom a solitary affair. It is often conducted by a team of scientists—in many cases several teams. For example, one recent physics journal article had 5,154 researchers listed as co-authors![7] Of course, this is an extreme case, but the fact is that many articles in scientific journals present the findings of collaborative works. According to a 2016 article in *The Economist*, as of 2015, the average number of authors on articles in science journals was 4.4.[8] What is more, even articles that have a single author likely depend on the work of other scientists to at least some degree, which explains all of the citations of previous work in the reference sections of these articles.

Although one might be tempted to think that collaboration in science is a recent development, it's not. Historian of science Kathryn M. Olesko has argued that even the most prominent cases of "solitary geniuses," Isaac Newton and Charles Darwin, were not really solitary.[9] They both relied on data collected by others as well as on the scientific work of others in developing their own theories. This does not lessen their achievements, but it does reveal that collaboration is not a new thing in science, and it is at the heart of some of the greatest discoveries in the history of science. To illustrate this point, consider the work of Gregor Mendel; he is usually portrayed as the father of genetics, a lonely pioneer working in isolation, who eventually discovered the material basis of heredity. It is exactly because of this isolation, the story goes, that his work published in 1866 did not become widely known until 1900 when three other naturalists independently reached the same conclusions he did. As a result, his priority was only acknowledged posthumously, and the world was deprived of his knowledge for 34 years. Sad story, right? Well, it would be if this was all there was to the situation. However, a closer look at what exactly Mendel wrote and the related historiography suggests that there is more to the story. Mendel was not trying to develop a general theory of heredity. He was rather trying to understand the development of hybrids and how heredity occurs in them from a practical point of view related to the agricultural context in which he was working. At the same time, a number of scholars including Charles Darwin, Francis Galton, August Weismann, Hugo de Vries, and several others attempted to develop theories of heredity contrary to what Mendel was doing. It was their work, and advancements in other domains such as microscopy and cytology, that eventually created a new context that allowed for a reappreciation of Mendel's experimental approach. Simply put, our understanding of heredity advanced due to the work of a scientific community, not of Mendel alone.[10]

Scientific progress relies in a significant way on collaboration among scientists. Of course, such collaborations require scientists to trust one another and the information that they share. The old adage that a chain is only as strong as its weakest link provides an apt analogy here. For a chain to be strong every link must be strong. Similarly, for a scientific claim (or any claim for that matter) to be epistemically certain, every bit of information that is relied on in supporting that claim must be epistemically certain. Is it really plausible that every single collaborator to a scientific project is in a position to provide epistemic certainty for every result and calculation they contribute to the overall findings? It seems not. If we cannot be certain that each collaborator is certain about all of their contributions, we cannot be certain about the final product of the collaboration. There cannot be any weak links if we are to achieve certainty.

Of course, as we emphasized in earlier chapters, just because we cannot be certain it does not mean that all claims are equally worthy of acceptance. Far from it! Some sources are reliable, and others are not. Fortunately, there are a number of social structures in place to aid in assessing the reliability of scientific claims. In order for a claim to become accepted by the scientific community it must undergo a peer-review process by experts in the same field who evaluate the questions asked, methods used, data collected, and conclusions drawn. It is because of these sorts of evaluation processes that science produces knowledge that is so well supported by the evidence. Nevertheless, nothing can guarantee certainty. Sometimes peer review fails to weed out results produced by poor methodology. Those conducting the peer review process are fallible, too, so it is always possible that a mistake may allow poor-quality research to slip through the cracks. Thus, despite the social structures in place, uncertainties of various kinds exist.

But, Doesn't the Scientific Method(s) Guarantee Certainty?

At this point, we might be tempted to appeal to the scientific method that many of us were taught in school. Doesn't this method make it so that the sorts of uncertainties we have been discussing do not infect science? Cannot we simply follow the scientific method and be guaranteed to reach the truth? Short answer: no. The main reason this is so is simply that such a thing as *the* scientific method does not exist. As historian of science Daniel P. Thurs has explained, the so-called scientific method (at least when understood to be some sort of universal, discipline neutral method) is a myth. Even an agreed upon account of what the purported method is does not exist. As Thurs noted, "even simplistic versions vary from three steps to eleven. Some start with hypothesis, others with observation. Some include imagination. Others confine themselves to facts."[11] If the very nature of the scientific method is itself uncertain, attempting to follow it is not going to yield certainty. After all, we cannot be certain that we have followed *the* scientific method when it is not certain exactly what this method is.

Let us assume that there is such a thing as *the* scientific method and there is no dispute about the steps of this method. Most representations of such a method agree that it involves making observations, formulating a hypothesis, testing the hypotheses via an experiment, and then, on the basis of the outcome of the experiment, accepting or rejecting the hypothesis.[12] Does following these steps produce certainty? It is worth considering each step.

First, consider making observations. You might think that we can be certain of our observations, can't we? One problem, as we explained in Chapter 1, is that our senses can deceive us. Beyond that, the "theory-ladenness" of observation also suggests that the answer is "no." In philosopher and historian Thomas Kuhn's terminology, scientists interpret observations they make through the lens of their particular "paradigm" ("beliefs, values, techniques, and so on shared by the members of a given community").[13] As a result, how we understand or interpret a particular observation depends on the particular paradigm with which we are working. We focus on particular aspects of what we observe because our paradigm deems them important. For example, as Kuhn explained, astronomers in Europe did not "see" changes in the heavens (such as the appearance of new stars in the night sky) until after the old view that the heavens are unchanging was abandoned for Copernicus's paradigm. This is in stark contrast to Chinese astronomers who observed the appearance of many new stars long before Europeans did. Why was this? The Chinese did not have better equipment or techniques. Presumably, they were not more studious observers either. It also was not because the stars were only visible in China. The difference was that the Chinese were not encumbered by a view of the heavens as unchanging. Rather, they were working under a different paradigm than the one guiding the pre-Copernican Europeans, one that prompted them to take note of the evidence for new stars.[14] The content of observations in science is at least in part determined by scientists' paradigms, and paradigms have changed throughout the history of science. Hence, there is uncertainty as to what exactly has been observed and how the observations are to be interpreted.

What about formulating a hypothesis? This seems pretty straightforward. Perhaps we can be certain that we have formulated the hypothesis we take ourselves to be formulating. Admittedly, we may not have come up with a true hypothesis or even a very good one. But, this is where testing comes into the picture. We test hypotheses via experiments and determine whether to accept the hypothesis or reject it. Uncertainty arises here. What happens when an experiment yields a result that is contrary to the hypothesis? Is the hypothesis rejected? It depends on whether or not the particular experimental result counts as a mere anomaly (something that does not fit well with the hypothesis but is not deemed important) or as evidence against the hypothesis. However, whether the result is considered an anomaly or evidence for rejecting the hypothesis depends on the way the result is interpreted. Is the fact that Mercury's orbit around the sun does not proceed as Newton's theory predicts a mere anomaly to be ignored or a reason to look to another theory? Prior to Einstein's work on relativity, Mercury's orbit was simply taken as an anomaly, but after Einstein proposed his theory that same result was taken to be evidence against Newton's theory.[15] Similarly, for results from controlled experiments: whether a particular result is an anomaly or a reason to reject a given hypothesis depends on how it is interpreted. And, of course, like observations, interpretations of an experimental result depend on the accepted paradigm employed by those evaluating the results. Again, we face uncertainty in how we are to interpret the results of experiments used to test the hypothesis.

What about accepting or rejecting a hypothesis on the basis of experimental results? This step is uncertain as well for at least two reasons. First, given our fallibility it is always possible that we make a mistake in our inference here. We might think that an experimental result warrants accepting or rejecting a hypothesis when it doesn't. Second, as Kuhn pointed out, the history of science shows us that scientists do not tend to actually treat experimental outcomes that conflict with their hypotheses as providing counters that demand the rejection of the hypotheses. Instead, such outcomes are treated as anomalies. And, in the face of anomalies, scientists tend to "devise numerous articulations and *ad hoc* modifications of their theory in order to eliminate any apparent conflict."[16] A clear example of this is the numerous epicycles that were added to the Ptolemaic model of the universe in order to make the theory fit with the observational data. Rather than abandon this model, a number of ad hoc modifications were tacked on to it so that it could yield the correct observational predictions. The prevalence of this sort of modifying may be part of the reason for Max Planck's cynicism that we noted earlier. It may be that "a new scientific truth does not triumph by convincing its opponents and making them see the light" because ad hoc modifications are the typical response to anomalies.[17] Whatever the case may be, it does not seem that scientists strictly follow this step in the purported scientific method at all. If the step is not even adhered to, it cannot provide certainty.

None of this is to say that science is purely subjective or relativistic. Rather, this is simply to point out that there is no scientific method that can give us certainty. We have already discussed how our ordinary thinking leaves us with uncertainty. And, as Albert Einstein said, "the whole of science is nothing more than a refinement of everyday thinking."[18] Scientific reasoning is the same sort that we employ in our everyday lives. As cognitive scientists Hugo Mercier and Dan Sperber explained, "Scientists' reasoning is not different in kind from that of lay people. Science doesn't work by recruiting a special breed of super-reasoners but by making the best of reasoning's strengths."[19] The way that science makes the best of reasoning's strengths is by implementing the sorts of social structures that we mentioned earlier: peer review, discussions, independent testing, debate, and so on. We explore more fully how scientists reason in the next chapter. For now, it is enough to recognize that their reasoning won't guarantee certainty.

But Doesn't Science Rely on Mathematics?

Although science does not attain certainty by way of the scientific method, one might think that it does so through mathematics. Many contemporary scientific theories are expressed as mathematical formulas. And when we want a clear example of reasoning that is precise and certain, mathematics is a good place to look. After all, what can be closer to certain than math?

Unfortunately, for those set on science being certain, appealing to mathematics will not do the trick. Mathematics is itself uncertain and does not provide us with certainty either. It is worth taking a bit of time to discuss one of the classic illustrations of uncertainty in mathematics. According to the January 22, 1978 issue of the *New York Times*, mathematician and logician Kurt Gödel was the "discoverer of the most significant mathematical truth of this [the twentieth] century."[20] This was Gödel's famous *Incompleteness Theorem*. The exact details of this theorem and the proof that demonstrates its truth are not important for our purposes. However, the upshot of this theorem definitely is. With his *Incompleteness Theorem*, Gödel showed—and it is pretty much universally accepted that he is correct—that some mathematical questions are undecidable: they cannot be shown to be true or shown to be false mathematically. More precisely, he established that "if there are no contradictions in mathematics, then there exist mathematical statements that can neither be proved nor disproved."[21] This means that mathematics is incomplete—there are mathematical claims whose truth or falsity will remain uncertain. As William Byers aptly pointed out, this constitutes "the existence of fundamental limits" to what we can know.[22] The presence of this sort of limitation in mathematics reveals that there will always be some uncertainty in mathematics. By extension, there will also be uncertainty when it comes to anything that is based on mathematics or formulated in mathematical terms.

Of course, as we have emphasized many times now, lack of certainty does not entail that something is unreliable or untrustworthy. Even though mathematics is not certain, it is still an incredibly reliable source of knowledge and very effective. As philosopher of mathematics Marcus Giaquinto noted, "we cannot be certain of the reliability, regarding finitary consequences, of much mathematics."[23] Nonetheless, "a high degree of confidence in the reliability of a large amount of classical mathematics has already been justified" by way of logical proofs.[24] This means that, in spite of disagreements and uncertainty, it is reasonable to have faith in mathematics. We just cannot have epistemic certainty.

Inherent Uncertainty

In some cases, uncertainty in science is so evident and profound that it is explicitly acknowledged, as in the case of the *Uncertainty Principle* put forward by physicist Werner Heisenberg:

> At the instant of time when the position is determined, that is, at the instant when the photon is scattered by the electron, the electron undergoes a discontinuous change in momentum. This change is the greater the smaller the wavelength of the light employed, i.e., the more exact the determination of the position. At the instant at which the position of the electron is known, its momentum therefore can be known only up to magnitudes which correspond to that discontinuous change; thus, the more precisely the position is determined, the less precisely the momentum is known, and conversely.[25]

Simply put, the idea is that the closer we are to certainty about where a particular electron is, the more uncertain we are about how fast it is moving and in what direction, and vice versa. The *Uncertainty Principle* entails that there are simply things that we cannot know about the natural world. Importantly, the limitation here is *fundamental*, just like the *Incompleteness Theorem* in mathematics. Improved technology, better theories, better techniques, etc. will not make a difference. We simply *cannot* know both the position and momentum of subatomic particles at the same time.

The *Uncertainty Principle* reveals an inherent limitation on what we can know, and it has important implications beyond just subatomic particles. Intuitively, if we cannot be sure of where an electron is, then even if we could be certain of how fast it is moving, we cannot be sure of where it will be in the future. Similarly, even if we could be certain of where an electron is at a particular time, if we cannot be sure of how fast it is moving (and in what direction), then we cannot be certain of where it will be in the future or its future velocity. This means that we are stuck with an irreducible uncertainty when it comes to predicting the future. (We discuss uncertainty in predictions more fully in Chapter 13). This sort of inescapable uncertainty further illuminates Carlo Rovelli's remark in the epigraph that thinking science is about certainty is the "deepest misunderstanding about science." We will always lack knowledge of some sort, and so we will always face uncertainty.

As we have been at pains to emphasize throughout our discussion, uncertainty in science does not denigrate science or scientists. Science produces some of our best-confirmed knowledge of the natural world. So what is the point? Why should we care that science is inherently uncertain? Rovelli got to the heart of things when he stated that "the core of science is not certainty, it's continual uncertainty."[26] Once we appreciate that *all* science is inherently uncertain, discovering uncertainties in the science of climate change, vaccination, human evolution, genetic testing, or forensic science should not incline us to think that they are inferior to other sciences. It is crucial to remember that being uncertain does not mean being unreliable or unreasonable. Let us now turn our attention in the next chapter to examining the uncertainties present in the explanations provided by our case studies (climate change, vaccination, human evolution, genetic testing, and forensic science) as well as by science in general.

12 | Uncertainty in Scientific Explanations

> To be accepted as a paradigm, a theory must seem better than its competitors, but it need not, and in fact never does, explain all the facts with which it can be confronted.
>
> —*Thomas S. Kuhn*[1]

The Goals of Science

Although there are many specific aims of science, three of its most important general goals are explaining, predicting, and intervening upon (to the degree that it is possible) natural phenomena. All of these are connected to *the* central aim of science: *understanding* natural phenomena. Our primary way of achieving understanding is by way of scientific explanations. In fact, some go so far as to argue that it is not possible to have scientific understanding without scientific explanations.[2] Once we possess understanding of natural phenomena via scientific explanations, we can use that understanding to predict and sometimes intervene to affect what will happen. (We discuss prediction and understanding more fully in the next two chapters.) The first step in this process is explaining the phenomena we observe—why things occur, how they occur, why they have the features they do, and so on.

As we noted in the previous chapter, the phenomena we seek to explain are extremely complex. Recall the figure representing causal influences on the occurrence of obesity from the previous chapter—as complex as that figure was, it failed (by a long shot!) to represent all of the causal influences on this phenomenon. Once we appreciate the complexity of the phenomena we want to explain, as well as the limitations of our perceptions and representations, we may end up wondering how it is possible for us to generate scientific explanations of anything at all. The way we do this lies at the heart of science—as philosopher and psychologist J. D. Trout observed "the complexity of these phenomena are untrackable without the prosthetics of models."[3] Roughly, the idea is this: we have scientific theories (which consist of laws, models, and principles) that offer broad generalizations about the world; we use the models that are part of those scientific theories, or we construct new models on the basis of scientific theories, to generate explanations of specific natural phenomena.

What exactly is a scientific model though? In simplest terms a *model* is a representation. A scientific model represents some aspect of the world in a way that makes the phenomena of interest more tractable. Philosopher Henk de Regt has argued that "the function of a model is to represent the target system in such a way that the theory can be applied to it."[4] An analogy may be helpful here. A city map represents specific aspects of a target system (the city), the locations of key landmarks and the city's streets, say. Of course, a city map does not represent every aspect of a particular city, but it does represent the relevant features for specific purposes such as navigating the city. Similarly, a model does not represent every aspect of a natural system. Instead, it represents certain aspects in a way that allows us to apply the appropriate theory and provide explanations of natural phenomena occurring in that system. This is one of the main ways that science makes contact with the world. As philosopher Ronald N. Giere explained "models are the primary (though not the only) representational tools in the sciences."[5]

A defining feature of scientific models (like models in general) is that they have to make simplifying assumptions—just like the map of a particular city. Sometimes these simplifications are brought about by way of approximation; other times, models simplify by way of assuming things that we know are false. In other words, the models used in science are *idealizations*. For example, when we want to explain the behavior of an actual gas we can do so by way of the Ideal Gas Law. This law helps us generate a model that describes the behavior of ideal gases. There is one small catch— there is no such thing as an ideal gas! Nonetheless, we use this idealization to model the behavior of actual gases. Such idealizations are not the exception in science but the norm. As we have already discussed, we are limited beings, and the phenomena we seek to explain are enormously complex. In light of these facts it is unsurprising that "idealizations are both rampant and unchecked in science."[6] We simply cannot do science without them. The use of idealizations leads to uncertainty in scientific explanations. After all, if the models (and often the theories that underlie these models) we use to generate explanations are idealizations, those explanations will be incomplete in various ways. Models cannot include every single aspect of a target system, and when they do include a particular aspect it is often only an approximation. This happens either because we lack the information or because we do not think it is relevant for explaining the phenomena of interest. For example, a city map often leaves out information concerning population demographics because such information isn't relevant to the goal of explaining how to travel throughout the city. Therefore, there is always information that is not included in a model, and thus every model is at best an incomplete representation of some aspect of the world. And where there is incompleteness, there is uncertainty.

What Is a Scientific Explanation?

Before delving more deeply into why scientific explanations are uncertain, it will be helpful to take a step back and think about what a scientific explanation is. A good place to start is with everyday explanations. We explain things all of the time. Why was your friend late to dinner? He had to work late and so missed the bus that would get him to dinner on time. Why did the dog eat the sandwich you left on the table? The dog was hungry and the sandwich was in reach. Why did the glass panes in the door break? Someone threw a stone at them. What do these explanations have in common? They describe dependence relations. More specifically, they describe how one thing causally depends on others. Your friend's working late caused him to miss the bus, which in turn caused him to be late to dinner. In other words, his being late to dinner was causally dependent on working late and missing the bus. The dog's being hungry was a cause of it eating the sandwich, and so the dog's eating the sandwich was dependent on its hunger. The stone striking the glass panes with sufficient force was a cause of the panes breaking, and so the panes' breaking was dependent on the stone striking them.

Scientific explanations are very similar to ordinary explanations—they also are a matter of describing dependence relations. In the majority of cases a scientific explanation is simply information about causal dependence relations.[7] Why do human greenhouse gas emissions count as an explanation of increased average global temperature? It is because human greenhouse gas emissions are contributing to increasing temperatures: when the concentration of greenhouse gases in the atmosphere increases, then the amount of energy absorbed by them and retained in the atmosphere also increases, and in turn the average temperature also increases. Simply put, when it comes to scientific explanations the aim is to identify causes and causal mechanisms. Broadly speaking, we have a scientific explanation of a phenomenon when we know the causes of that phenomenon. Our scientific explanation will be more complete the more information it provides about the nature of the causes (and the causes of those causes) and about how those causes produce the phenomenon.

How Is Scientific Explanation Related to Theory?

Now that we have briefly discussed the nature of scientific explanations, let us consider how they are related to scientific theories. Scientific theories tend to be broad generalizations that cover a wide range of phenomena, such as general relativity, quantum mechanics, evolution. Importantly, calling something a scientific "theory" does not imply that it's not known to be true or that it is not well supported! Whereas the word "theory" is often used in everyday situations to denote a guess, in science "theory" refers to a robust account that provides understanding. The three theories just mentioned are some of the best-confirmed ones in science. Contrary to what some people claim, evolution is not *just a theory* in the vernacular sense. Indeed, it is a theory in the scientific sense. So, evolution is just a theory—it's just like general relativity and quantum mechanics: a collection of models and principles that can be applied to explain the available evidence from natural phenomena.

One might wonder though—if the scientific theories we just mentioned are so well supported, why aren't they considered laws? Although this sort of concern is quite commonplace, it rests on a misunderstanding of the nature of scientific theories and laws. Theories do not become laws when they receive enough support. Even though at times both theories and laws can be used to generate explanations, generally they serve different purposes. Roughly, scientific laws tend to report on observations and offer a way of predicting what will happen, whereas scientific theories offer broad explanations of why particular things happened.[8] This is, of course, a bit of an oversimplification because, as we pointed out earlier, scientific theories often include laws. The most important point for our purposes, though, is that scientific laws are not in any sense more certain than theories, and, furthermore, they are not certain in any absolute sense. Rather, scientific laws are uncertain because they are idealizations. For instance, the Ideal Gas Law describes the behavior of gases whose molecules are frictionless, are not subject to intermolecular attraction, and are dimensionless. There are no such gases though. Philosopher Nancy Cartwright has argued that the most fundamental scientific laws in physics are strictly speaking *false* because they do not accurately describe actual phenomena.[9]

For instance, Newton's law of universal gravitation includes the idealizing assumption that only gravity, and no other force, affects the bodies under investigation. But, of course, this is never the case. There is always more than just one force affecting a physical body. So, strictly speaking, Newton's law does not apply to any real objects. Therefore, when we use the law to predict the behavior of objects, we are not able to predict with certainty. Newton's law of universal gravitation (like other fundamental laws) does not apply perfectly to actual phenomena because it is only an approximation. Again, this does not mean that scientific laws aren't well supported by the available evidence (they are) or that we cannot use them to come to understand the world (we can); it simply means that even scientific laws and the predictions they provide are not certain. The same is true of the explanations and predictions provided by scientific theories. Hence, the fact that something is a scientific theory rather than a scientific law in no way implies that it is not worthy of accepting as true.

So how are scientific explanations related to theories? It is through models. Theories contain models, or are used to construct models, and those models are used to generate explanations of specific natural phenomena. Once we have constructed models and generated explanations, we must determine which explanation to accept. As the epigraph from Thomas Kuhn indicates, theories and the explanations generated by them never explain all of the facts so we have to compare theories and the models that they can be used to generate. This is done by way of comparing the different explanations of natural phenomena that competing models and theories provide. When we do this, we are engaging in what is described as *inference to the best explanation*. This is a method of inference that involves inferring that the best sufficiently good explanation among the available possible explanations of some phenomenon is true (or at least likely to be true). Let us take a look at a very simple example of this kind of reasoning. You come home and notice several bags full of groceries on your kitchen counter. Here are a few potential explanations for the grocery bags being on your kitchen counter:

1. Your spouse went grocery shopping recently and put the bags there.
2. Members of a charitable organization, such as the Red Cross, have been randomly giving groceries to people in your neighborhood, and you were randomly selected.
3. The bags of groceries were placed there by a burglar who broke into your house in order to leave you free groceries.

Obviously, the spouse explanation is a much better explanation of the grocery bags you find on your kitchen counter than is the Red Cross explanation. And the benevolent burglar explanation is far worse than the Red Cross explanation. Why? The spouse explanation is more *virtuous* than its rivals. First, it is simpler than the others because it does not involve appealing to exceedingly strange activities by the Red Cross and your random selection, nor does it refer to a benevolent burglar, the existence of which is exceedingly rare if there are any at all. Second it fits better than the other explanations with your background information about your spouse, the sorts of things the Red Cross does, the safety of your neighborhood, and so on. It also explains both the groceries and why you have not heard anything about this strange initiative of the Red Cross as well as why there does not seem to be any signs of a break-in. The other potential explanations do not explain these facts. For these reasons (and probably more), the spouse explanation is both a very good explanation of the bags of groceries you found on your counter, and it is better than all of its rivals. Inferring to the best explanation yields that you should believe that the spouse explanation is true.

Of course, the grocery example is simplistic and it may fail to look like scientific practice when we first think about it. But the inference described here exhibits the same sort of reasoning that we encounter in science. Scientists evaluate competing explanations in terms of their explanatory virtues. Scientists prefer explanations that are simple, fit with background information, and have a good amount of explanatory power (explain various different instances of a phenomenon and/or explain a large number of details concerning the phenomenon). Scientists are continually in the process of inferring to the best explanation. As Kuhn said "no theory ever solves all the puzzles with which it is confronted," so rather than simply looking at one theory and the explanations it offers in isolation "it makes a great deal of sense to ask which of two actual and competing theories fits the facts *better*."[10] Not only does this make sense, but it is also what we find in science.

Consider a classic example of inference to the best explanation in science: the discovery of Neptune. It was discovered because the best explanation of why Uranus's orbit did not match what was predicted by Newton's laws was that there was another as yet undiscovered planet interfering with the orbit of Uranus. Not long after astronomers postulated this unknown planet because it best explained Uranus's orbit, Neptune was discovered through empirical observation. This particular inference to the best explanation led to an exceptional discovery, but there is nothing exceptional about the use of inference to the best explanation in science. The electron was discovered as a result of this sort of inference, and so was the process of photosynthesis. The heliocentric model of the solar system was adopted because it offered a better explanation of the available data than the geocentric model. General relativity rather than ether theory is accepted because the former best explains the evidence. And on and on.[11] As philosopher Clark Glymour remarked, "one can find such arguments in sociology, in psychometrics, in chemistry and astronomy, in the time of Copernicus, and in the most recent of our scientific journals."[12] In fact, J. D. Trout has argued that the success of modern science boils down to two things: hitting on theories that were on the right track (they were approximately true) and the use of inference to the best explanation.[13]

Interestingly, what we find when thinking about the relationship between explanations and theories is that they are mutually supporting. We infer that the best explanation of some phenomena is true. The truth of the particular explanation provides support for thinking that the theory used to generate the model it comes from is true. The more likely a theory is to be true, the more likely the models and explanations that are produced from that theory are to be true. This does not mean that a theory that has continually produced the best explanations is guaranteed to always produce the best explanations or that the theory is certain. It does mean that theories can provide support for the models and explanations derived from them and vice versa.

All Scientific Explanations Are Uncertain

We saw in earlier chapters that the science surrounding climate change, vaccinations, human evolution, genetic testing, and forensic science produces scientific explanations that are uncertain. Having already argued that the scientific explanations produced in these domains are not inferior to other scientific explanations, let us now consider why *all* scientific explanations are uncertain. Whether or not scientists working with a particular theory face the exact same challenges, all science must deal with the two difficulties of extreme complexity and human limitations. In the face of this complexity we have no choice but to rely on models; crucially, this involves idealization and approximation.[14] The use of idealizations and approximations in science is absolutely necessary given our limitations and the complexity we face, and this entails that the explanations produced in any area of science are uncertain.[15]

Another source of uncertainty in scientific explanations is that we are always working with limited information. We have limited, indirect epistemic access to the past (hence, for example, some of the uncertainty present in explanations concerning human evolution). We are also limited in our access to information in the here and now. Recall the *Uncertainty Principle* from the previous chapter. It is simply a fundamental feature of the universe that we cannot know the location and the momentum of a particular electron at a given time. Without all of the information, which we never have, a scientific explanation can never achieve absolute certainty.

Finally, it is worth thinking about an uncertainty that we face whenever we determine whether a given theory or explanation is correct. As Kuhn has aptly noted, "the act of judgment that leads scientists to reject a previously accepted theory is always based on more than a comparison of that theory with the world"; it involves comparing the theory with both the world and with other theories.[16] We are always inferring the truth of the best *available* explanation. When we do this, however, we face what philosopher P. Kyle Stanford has called the "*problem of unconceived alternatives*."[17] This is the problem we face because we cannot be certain that we have thought of all of the best potential explanations of a given phenomenon. There may exist other really good explanations. If we cannot be certain that we have considered all of the best potential explanations of a phenomenon, then we cannot be certain that the explanation that we have judged as the best of those available is really the correct explanation; there may be a better one that we have simply not thought of yet.

That no explanation produced in science is absolutely certain is important to recognize for at least two reasons. The first is that once we see that no scientific explanation is certain, it is clear that the fact that there are uncertainties in the explanations provided in some domain, such as climate science, does not give us any good reason to dismiss the science or those explanations. The second reason is that appreciating that this is the nature of scientific explanations helps us to better understand them and science more generally. As we have emphasized before, we cannot have certainty, but we can have scientific explanations that are extremely well-supported by the evidence and worthy of accepting as true. The fact that no scientific explanation is certain means that we should always be open to the discovery of new evidence and better explanations. At the same time, we should believe according to our evidence—when an explanation is very well-supported by the evidence, despite being uncertain, it is rational to accept it.

Let us turn our attention in the next chapter to uncertainties in scientific predictions.

13 | Uncertainty in Scientific Predictions

> Consideration of black holes suggests, not only that God does play dice, but that he sometimes confuses us by throwing them where they can't be seen.
>
> —*Stephen Hawking*[1]

Anticipating the Future

We all look to the future with some anticipation. One of the reasons that uncertainty can be stressful is that it makes it hard to plan for what is to come. After all, if we were certain about what will happen, life would be much easier! No one would get on the plane if they were certain that one of its engines would fail midflight. There would be no need to go to the store if one were certain that the store does not have the particular item that one needs. Deciding whether to pay for insurance would be an easy matter if we could predict the future with certainty. After all, why pay for travel insurance if you are certain that everything will go smoothly on your trip? Similarly for any other kind of insurance; you do not need it if you are certain that the bad event for which you are insured is not going to occur. Whether it is about buying insurance, bringing a jacket to the theater because we predict that it will be cold like it was last time we were there, or stopping by the store because we predict that they will have the ingredients we need to make our favorite dish, we make many predictions. And these predictions that we make are quite often reasonable, and many times they are accurate. But we are never certain about future outcomes. The same is true in science.

Before delving into the nature of scientific predictions and their uncertainties, we should consider two fairly common mistakes. The first is taking scientific predictions to be certain, and the second is thinking that because a prediction is uncertain it is not trustworthy. A prime example of the first mistake arises in connection with genetic testing. Often people tend to accept what is known as *genetic determinism*. The idea that an individual's genes fix the individual's characteristics so completely that environmental factors have little to no effect.[2] This, of course, is not correct. A variety of factors affect whether or not an individual displays a particular characteristic. So we cannot predict with certainty that since one has a particular gene or set of genes, one will also have a particular characteristic. The second mistake is evident in discussions of climate change. It is common to hear that since there is uncertainty about the role of human actions in climate change, we do not need to take steps to change our behaviors by doing things like lowering our greenhouse gas emissions. It may be that, as philosopher Lorraine Code has argued, some people use this uncertainty concerning the details to avoid taking any action or responsibility. This is so even when the larger issue, such as human actions affecting climate change, is very firmly established.[3] Whatever the case may be, it is surely a mistake to assume that because a prediction is uncertain, it is not worth accepting or taking seriously. It is not certain that you will lose the lottery when you buy a ticket, but it is a very good prediction given the extremely low odds of winning!

The Relationship Between Scientific Explanation and Prediction

A first step in better understanding the nature of scientific predictions is to consider how they are related to scientific explanations. A straightforward, but mistaken, picture of the relationship between scientific explanations and scientific predictions is that they are actually the same thing. One of the most prominent twentieth-century philosophers, Carl Hempel, held this view.[4] According to Hempel, scientific explanation and scientific prediction are essentially the same—the only difference is that explanation looks back at what has already happened, whereas prediction looks forward to what will happen. On Hempel's view, a scientific explanation is a deductive argument in which one uses general scientific laws and particular facts to deduce the occurrence of the phenomenon that is being explained. The only difference between a scientific explanation and a scientific prediction on this view is that, with a scientific explanation, we already know the phenomenon being explained has occurred, whereas with a scientific prediction we do not. For instance, we can explain why a particular glass in a door broke by deducing its breaking from general laws of force and momentum along with particular facts about the stone that was thrown at the door. We can similarly predict that a particular, as of yet, unbroken glass in a door will break by deducing this conclusion from the same general laws of force and momentum along with particular facts about how a stone is being thrown at it.

Despite the simple picture of the relationship between scientific explanations and scientific predictions that Hempel's view provides, it is widely acknowledged to be false. The problem is that it seems obvious that scientific explanations and scientific predictions come apart. More specifically, it is clear that one can have a good scientific prediction of a phenomenon without having a scientific explanation of it. The classic illustration of this fact, and a general problem with Hempel's account of scientific explanation, involves a flagpole on a sunny day. If you know the height of a flagpole, say 20 feet, you can both predict and explain the length of its shadow at a particular time of the day by also considering the angle of the sun, the rules of trigonometry, and the fact that light travels in a straight line. So far so good. This is a case in which explanation and prediction are not that different. The length of the shadow can both be predicted and explained. However, if you know the length of the shadow that is cast by the flagpole along with the angle of the sun, using trigonometry and the fact that light travels in a straight line, you can accurately predict the height of the flagpole before you measure it, but you cannot explain why the flagpole is 20 feet tall. The explanation for why the flagpole is 20 feet tall has nothing to do with the angle of the sun's rays or the shadow it casts on the ground. The reason the flagpole is 20 feet tall has to do with the purposes of the people who placed the flagpole where it is, the materials they had to work with, and so on. Thus, it seems that, in this case, we have a good scientific prediction without a scientific explanation—the two are not just the same thing looking in different directions, as Hempel thought.[5]

Nonetheless, scientific explanations and scientific predictions are closely related. As philosopher Heather Douglas nicely put the point "the relation between explanation and prediction is a tight, functional one: explanations provide the cognitive path to predictions, which then serve to test and refine explanations."[6] Scientific explanations are often our path to scientific under-standing, which allows us to make scientific predictions. When we check to see if those scientific predictions are accurate, we either confirm or disconfirm the scientific explanations used to make the predictions. This, in turn, allows us to improve our scientific un-derstanding of the phenomenon in question and, as a result, make better predictions. Indeed, some predictions might go so far as to question our explanations were they to be confirmed. For example, legend has it that John Burdon Sanderson Haldane, an important contributor to twentieth-century evolutionary biology and ge-netics, once stated that he would give up his belief in evolution if someone found a fossil rabbit in the Precambrian. What Haldane did was actually predict that nobody would find a mammalian fossil in rocks older than the common ancestors of all vertebrates, which includes the Precambrian rocks, according to evolutionary theory. Similarly, since evolutionary theory suggests that the ex-tinction of non-avian dinosaurs occurred about 65 million years ago and the evolution of our species occurred about 200,000 years ago, we can predict that we would not find the fossil skeleton of Fred Flintstone next to the fossil skeleton of Dino, the family's pet dinosaur, or, more generally, we would not find the fossil of a mammal next to a dinosaur fossil. If any of these predictions were confirmed, it would take down evolutionary theory. But this is not the case, and exactly because such predictions are not confirmed, our confidence in the explanations provided by evolutionary theory continues to increase.

In spite of this close relationship, scientific explanations and scientific predictions do not always work hand in hand. In fact, the models used to facilitate scientific explanation and scientific prediction are not always the same. In other words, sometimes the most explanatory scientific model does not provide the most accurate predictions, and sometimes the scientific model that produces the most accurate predictions is not the most explanatory.[7] Why is this? It is because when constructing either scientific explanations or scientific predictions we face the same two issues: extreme complexity of the target phenomena and human limitations. Scientific models help us to navigate these issues, but, as we have already noted in earlier chapters, they never do so perfectly. As a result, some models are better suited to the aim of explaining; others are better suited to generating predictions. It is because of our limitations and the complexity of what we are trying to understand that we cannot always generate the best scientific explanations and the best scientific predictions by using the same models. To help illustrate this point, recall from Chapter 6 that one way to help generate accurate predictions concerning climate change is to apply robustness analysis to a number of different climate change models (often these models rely on conflicting assumptions). The result is a kind of meta-model that is made up of the various climate change models. Often, robustness analysis can provide us with accurate predictions of how climate change will proceed—more accurate than the individual models used. However, it does not help explain climate change. After all, it is constructed out of a number of other models that are not consistent with one another. Thus, the best predictive model is not very explanatory at all. Hence, although scientific explanations and scientific predictions often work in concert, they do not always do so.

All Scientific Predictions Are Uncertain

In the previous chapter we discussed why all scientific explanations are uncertain. The purpose of that was twofold. It helped clarify the nature of scientific explanations, and it made clear that the science surrounding climate change, vaccinations, human evolution, genetic testing, and forensic science is not inferior to other domains of science. A similar point is true when it comes to scientific predictions. The scientific predictions that one finds in connection with climate change, vaccinations, human evolution, genetic testing, and forensic science are not certain. But, neither are the predictions made in any other area of science. Consequently, uncertainty in their predictions is not a reason to doubt the reasonableness or the legitimacy of the respective science.

To understand why scientific predictions are uncertain, it is helpful to first consider what would have to be the case in order to have a scientific prediction that actually was certain. In order for a scientific prediction to be certain we need to (at least) be (1) certain that a specific regularity (scientific law) holds without exception and (2) certain that the phenomenon being predicted falls under that regularity. Here is a simple illustration. In order for the prediction that the next bird we see will be black to be certain, we need to be certain that, for example, (1) all crows are black and (2) the next bird we see will be a crow. Of course, even in this simple case we cannot achieve certainty. First of all, we cannot be certain that the next bird we see will be a crow—especially if we are simply observing birds in the wild. Second, although we might generally accept that all crows are black, this regularity is not without exception. There is a small number of naturally occurring white crows.[8]

It might be tempting to think that the extreme difficulty in finding regularities without exceptions only arises for mundane regularities like "all crows are black." Such regularities are imprecise, but scientific laws are expressed as mathematical formulas and are therefore expected to be very precise. While this is true, it does not mean that scientific laws hold without exception. Consider Newton's law of universal gravitation. This law describes the attraction between two bodies with a concise mathematical formula: $F = G(m_1 m_2)/R^2$. Without going into the specifics of Newton's law, the gist is that two bodies attract each other with a force that is described precisely by this formula. Does this law hold without exception? Despite the fact that it is called the law of *universal* gravitation, the answer is "no." This law accurately describes the gravitational force between two bodies *only if* we assume that no other force is acting upon either body. Why is this an issue? Because in any actual situation, such an assumption is false since many different forces are acting on any given body. Take the gravitational attraction between Earth and the sun, for example. Newton's law tells us that Earth is attracted to the sun with a particular force. However, the resulting force is not completely accurate because the sun is not the only object attracting Earth. Every other planet in the solar system (and every other object) is exerting gravitational force on Earth as well. As a result, calculating the attraction between two bodies using Newton's formula always ends up being inaccurate. Of course, it is an extremely good approximation in most cases because the forces it does not take account of are very minor and so their effects are insignificant, but, strictly speaking, the law gives the wrong answer. This means that Newton's law of universal gravitation does not hold without exception—in fact, every actual situation is an exception! Recognition that this is true of every scientific law has led some to conclude that scientific laws should not be understood to be regularities without exceptions, but instead regularities that only hold in some circumstances.[9] Others have concluded that there simply are no genuine scientific laws.[10] Either way, scientific laws will not allow us to make scientific predictions that are certain because they are not without exception.

Even if, contrary to what we saw earlier, we assume that scientific laws hold without exception, this still does not seem to be enough to yield scientific predictions that are certain. The reason for this is that, as Stephen Hawking said, "God does play dice." When he made the statement in the epigraph to this chapter, Hawking was pointing out that our current best theory of fundamental physics—quantum mechanics—includes scientific laws that are probabilistic. Such probabilistic scientific laws do not tell us what exactly will happen in a particular situation. Instead, they tell us what the probabilities are that something will happen in a particular situation. So, for example, rather than telling us that a particular atom of a radioactive substance will decay within a 5-year period, probabilistic scientific laws only tell us that it is 75% likely that it will decay. As a result, even if we are absolutely certain that a particular probabilistic scientific law applies to the situation we want a prediction about, we cannot be certain of the future outcome. We only know the probability of various outcomes. The point that Hawking made, a point about which many philosophers and scientists agree, is that the universe is not deterministic.[11] The future is not fixed by scientific laws combined with what is occurring now (or what has occurred in the past). Instead, there are various ways that things could turn out, and this is why scientific predictions cannot be certain.

Let us set aside the issue of determinism by assuming that the universe is in fact deterministic. In this case, the scientific laws along with how things currently are will determine exactly how everything will be in the future. Would the truth of determinism allow our scientific predictions to be certain? The short answer is "no," because it is not enough that the universe actually is deterministic. In order to have scientific predictions that are certain, we need to be certain that we have correctly identified the scientific laws that govern the universe, and we also have to be certain that the particular situation that we want to make predictions about falls under the relevant law. Recall our discussions from earlier chapters about the enormous amount of complexity we face in describing the universe. Even a single phenomenon such as obesity is the result of a huge number of causal influences. In light of this, it is plausible that, even if the universe is deterministic, we may not be in a position to figure out what the exceptionless scientific laws really are. As Hawking said, perhaps the dice are simply thrown "where they can't be seen." But let us assume not only that the universe is deterministic, but also that we can be certain that we have discovered a scientific law that holds without exception. This is not enough to yield scientific predictions that are certain. We still need to be certain about the conditions we are applying that law to when formulating predictions. As mentioned earlier, even if we were certain that all crows are black is a universal law, this would not be enough to provide a certain prediction that the next bird we see will be black unless we were also certain that the next bird we will see will be a crow.

Can we be certain about the specific conditions we are examining when making predictions? Unfortunately, we have already seen in Chapter 11 that the answer here is also "no." According to Heisenberg's *Uncertainty Principle*, the closer we are to certain of where an electron is, the more uncertain we are of how fast and in what direction it is moving, and vice versa. Hence, even if we were certain that an electron is at a particular position now, we cannot be certain of where it will be 2 seconds later because we cannot know all the relevant information (in this case, we cannot be certain of the speed and direction that the electron is moving). Here is an analogous situation. Assume we know with certainty that the President of the United States is in Washington, DC, right now and that he will soon be traveling. Where will he be in an hour? We obviously cannot be certain because we do not know which direction the President is traveling or how fast. Similarly, if we are certain that the President is traveling due north at 60 mph, we cannot be certain of where the President will be in an hour without being certain of where he started. If we cannot know where the President started while also knowing how fast and in which direction he is traveling, we cannot predict with certainty where he will be in the future. When it comes to every single electron in the universe, we are in this situation: we cannot know both its position and the speed at which it is traveling, so we cannot be certain of where any electron will end up.

It might be tempting to think that small details about exactly where specific electrons are will not make a difference to predictions that are more macroscopic in size. Perhaps this is true, but minute differences in starting conditions can lead to enormous differences in outcomes. Take the weather for instance. One of the reasons that weather forecasting is so difficult is that incredibly small differences in starting conditions can yield large differences in the weather. As mathematician and physicist Roger Penrose explained, "Weather patterns have the property that they are 'chaotic systems,' in the sense that any particular pattern which develops will depend critically on the tiniest details of what happened before.... Indeed, it is probable that, in a month, say, tiny quantum effects will become so magnified that the entire pattern of weather on the planet would depend upon them."[12] Sometimes this is popularly referred to as the "butterfly effect"—the idea that something as seemingly insignificant as a butterfly flapping its wings at a particular time can lead to enormous changes in the weather, such as the occurrence of a hurricane on the other side of the planet. The point is simply that small differences in initial conditions can lead to very different future outcomes. Consequently, in order for a prediction to be certain, we would have to be certain that the conditions we are starting with are precisely what we take them to be.

Is It Pointless to Make Predictions?

As we have noted, in order for a scientific prediction to be certain, several conditions would have to be satisfied. First, it would have to be certain that the universe is deterministic: this is not certain.

Second, we would have to be certain that we have identified the actual scientific laws governing the deterministic universe: we are not. Third, we would have to be certain of the precise initial conditions that we are drawing our predictions from: we cannot be. Therefore, scientific predictions are always uncertain.

Things might seem pretty dismal when it comes to scientific predictions, but they are not. It is important to keep in mind that just because scientific predictions are not certain, it does not mean that they are not reasonable to accept, or even that we cannot know that they are true. Again, a lack of certainty does not mean that anything goes. Scientific predictions are extremely reliable, and we would be making a serious mistake in betting against them. After all, if you buy a lottery ticket for the Mega Millions lottery, is it certain that you will lose? No, but given the 302.6 million to 1 odds that you will lose, it would be foolish for you to quit your job because you bought a ticket![13] If you flip a fair coin 100 times, is it certain it will not land heads on any of the flips? No, but it would be foolish to bet that it will not. Similarly, if the current best science predicts that if we do not curb our greenhouse gas emissions there will be significant consequences, is it certain? No, but we would all be foolish not to accept this prediction and act accordingly.

When it comes to predicting which one of several different possible outcomes will actually materialize, there is uncertainty. However, because the different outcomes may be more or less likely to occur under certain conditions, we can estimate how likely a particular outcome is. Therefore, even though we may not be able to definitely predict whether outcome A or B will occur, we may be able to predict that outcome A is more likely than B, and this can be a valuable prediction. For example, whether or not a mistake will occur in the cell division in a woman's ovum is uncertain, but, if it occurs, it might result in an embryo with trisomy 21 (known as "Down syndrome"). Whereas we cannot accurately predict what will happen in each individual ovum, we do know that such mistakes are more likely to happen in women over age 40 than in younger women. Therefore, whereas a certain prediction about each outcome is not possible, we can reasonably predict the probability of such an outcome based on the available evidence, and thus we can predict that this is more likely to occur in older rather than younger women.[14] Predictions thus cannot be certain, but can nevertheless be very useful.

14 | Understanding Versus Being Certain

> What are the peculiar, special traits of our scientific world-picture? About one of these fundamental features there can be no doubt. It is the hypothesis that *the display of Nature can be understood.*
>
> —*Erwin Schrödinger*[1]

The Primary Aim of Science

There are many aims of science. We have already discussed two of those aims: to explain and to predict natural phenomena. Science also aims to provide us with new medicine and technology. It aims to make our lives better. Despite its numerous aims, when we are talking about the *epistemic aims* of science—those goals having to do with our cognitive grasp of the nature and structure of natural phenomena—the primary aim is understanding. According to Nobel Prize laureate P. W. Bridgman, "The act of understanding is at the heart of all scientific activity."[2] It is apparent from the epigraph that Erwin Schrödinger agreed. Such a view is also widespread among philosophers of science. Philosopher Henk de Regt has argued that "most philosophers agree on the idea that understanding—whatever its precise nature—is a central aim of science."[3] Other philosophers have not just referred to understanding as "a central aim of science" but as its ultimate epistemic aim.[4]

It is not difficult to see why philosophers and scientists agree about the importance of understanding. In the previous two chapters we discussed explanation and prediction, two very important goals of science that tend to be closely related to understanding. Why do we want explanations of natural phenomena? Because we want to understand why things occur and how they do so. Once we understand why things occur and how they do so, we can make accurate predictions about what will occur in the future. The better we understand natural phenomena, the more predictions we can make and the more accurate those predictions will be. Science aims at understanding, not certainty. Recognizing this fact allows one to not only better understand the nature of science but also its successes. Furthermore, it can help one avoid being led astray by (mis)representations of the uncertainties inherent in various scientific domains.

What Is Scientific Understanding?

To better appreciate that science aims at understanding rather than certainty it will be helpful to get clearer on what exactly understanding is. But, before that, let us briefly remind ourselves of what certainty is. As we discussed in earlier chapters, there are two kinds of certainty: epistemic and psychological. *Epistemic certainty* concerns the strength of one's evidence—one is epistemically certain when one's evidence makes it so that it impossible to be wrong about some claim. *Psychological certainty* is about being completely convinced of the truth of some claim—one has absolutely no doubt about the matter. We have already seen that while we might be psychologically certain about various claims, we cannot have epistemic certainty in science.

What about understanding? There are different kinds of understanding. Sometimes we use the term "understand" as a way of hedging, such as when one says, "I understand that you are upset with me." This use is a way of not fully committing to the claim that the other person is upset. Another common way that the word "understand" is used is simply as a synonym for "knowledge." "She understands that the test is today" is just another way of saying that "She knows that the test is today." When it comes to science, the sort of understanding we are concerned with is understanding of why and how. Why is it that plants cannot live without sunlight? How do plants use sunlight in the process of photosynthesis? And so on. Scientific understanding is the understanding of natural phenomena. So, when we say that someone "understands" photosynthesis, we are not hedging and we are not simply saying that she knows something. We are attributing a particular cognitive achievement to her; she grasps why and how photosynthesis occurs. In other words, she possesses a correct explanation of photosynthesis.

Let us explore scientific understanding in more detail. First, there is a difference between having a feeling of understanding and genuinely understanding something. In many cases a feeling of understanding is primarily the result of various cognitive biases such as overconfidence bias (discussed in Chapter 1) or hindsight bias (erroneously thinking after something has occurred that we could have predicted it all along).[5] Importantly, we can experience this sort of feeling of understanding without genuinely understanding a particular phenomenon. For example, someone might have read her daily horoscope and, as a result, feel like she understands why she is having a tough time at work—today is just not a good day for Virgos; the stars are out of alignment. It does not matter how strong her feeling of understanding is in this case. This person does not understand why things are rough for her today because astrology does not provide genuine understanding of why her life (or anyone else's) goes the way it does. Now, as philosopher Catherine Z. Elgin has pointed out, it might be true to say of this person that she understands astrology. In other words, this person does not just have a feeling that she understands astrology; she genuinely understands it. But this just means that the person "knows how [astrology's] contentions hang together, and is adept at reasoning within the framework that they constitute." The person does not understand the phenomenon she is experiencing—her tough day at work—because astrology "affords no understanding of the cosmic order."[6] Astrology does not provide an accurate account of the phenomenon in question, so it does not provide genuine understanding even if some people may feel that they understand why things happen because of their knowledge of astrology.

If understanding is more than a feeling, what exactly is it? Perhaps it is simply knowledge. It may seem plausible that understanding merely amounts to knowing various things—the more one knows, the deeper is one's understanding. Even if the person who subscribes to astrology feels that she understands why she is having a tough time at work, she does not really understand it because astrology is not true. Consider another case. Someone might know geometry without understanding it. It is possible that when it comes to geometry, one "knows all the axioms, all the major theorems, and their derivations" but only by way of memorization.[7] A teacher has told her that all of these things are true, and she has simply committed them to memory. In such a case we cannot say that this person understands geometry despite all her knowledge of the subject. The person who genuinely understands geometry does not simply memorize axioms and theorems. When someone understands geometry she "can reason geometrically about new problems, apply geometrical insights in different areas, assess the limits of geometrical reasoning for the task at hand, and so forth."[8] In such cases, it appears that understanding requires more than simply knowing various facts. Consequently, it seems that equating understanding with knowledge is a mistake.

But what more than knowledge is required for understanding? Philosophers tend to agree that understanding requires a kind of cognitive "grasping" that knowledge does not. According to philosopher Michael Strevens, scientific understanding is "a matter of grasping, roughly, the causes of phenomenon to be explained, and the facts in virtue of which they are causes."[9] Similarly, philosopher Stephen Grimm has argued that understanding "seems to require not just holding several related things before the mind, but in some way 'seeing' or 'grasping' or appreciating this relatedness."[10] Perhaps the best way to put this point is that understanding requires being able to use the information that one has. The person who understands geometry does not merely know a bunch of geometrical facts; she also grasps how these facts fit together and she is able to use the information in a variety of ways. Hence, when it comes to scientific understanding we might follow philosopher Finnur Dellsén and accept that scientific understanding is in some sense a matter of "grasping how to correctly explain and predict aspects of a given target" (in this case, the target will be some range of natural phenomena).[11] This does not commit us to claiming that someone has to be able to come up with new explanations or predictions in order to have understanding. Even if one's ability is limited to comprehending new geometrical explanations and proofs that others show her without being able to come up with new ones on her own, it still seems that she has something that the person who only memorizes geometrical facts lacks. So we should construe Dellsén's point broadly to include not only producing explanations and predictions but also appreciating the explanations and predictions of others.

A very helpful way of putting all of these insights on understanding together is provided by philosopher Henk de Regt.[12] According to him, when it comes to scientific understanding there are actually three senses of understanding that need to be distinguished: the phenomenology of understanding, understanding a theory, and understanding a phenomenon. The phenomenology of understanding is the felt sense of understanding that we discussed earlier. This is the sort of sense that we have when we take ourselves to understand something. Unfortunately, as we have noted, this sense may be the result of biases, so it can be misleading. Of course, it is nice when we have the phenomenology of understanding along with the understanding of a phenomenon, but such feelings can be had without genuine understanding, and, perhaps, genuine understanding can be had without such feelings. Understanding a theory is a matter of knowing the theory and being able to use it. In the earlier geometry example, the person who has simply memorized geometrical facts only has the first component of this sort of understanding. The reason she does not count as understanding geometry is that she lacks the ability to use it, and, moreover, she fails to grasp how the various things she knows about geometry are related. Finally, there is understanding a phenomenon, which is present when one has an adequate explanation of why and how the phenomenon occurred. It is this last sense of understanding at which science aims. Scientific understanding arises when one understands a theory and then is able to use that understanding to appreciate or generate adequate explanations and predictions of phenomena.

Before moving on to our discussion of why science aims at understanding rather than certainty, it is worth emphasizing one additional point about scientific understanding: it comes in degrees. In Dellsén's terms there is "complete understanding" of a phenomenon, which requires "one to grasp how to correctly explain and predict every aspect" of the phenomenon in question. But there is also "partial understanding," which requires "grasping how to correctly explain or predict *some* aspects" of the phenomenon.[13] The more of those aspects one can correctly explain or predict, the better one understands the phenomenon. This tracks quite well with our natural inclination to talk about some people better understanding something or having a deeper understanding than others. We might truly say of both an introductory biology student and a professional botanist that they understand photosynthesis. Yet we would also correctly say that the botanist has a deeper understanding of photosynthesis than the student has. This will be important to keep in mind as we turn toward discussing the success of science.

The Success of Science

Science and its many discoveries are among the greatest successes that humans have ever achieved. But what is the reason for this success? To answer this question, we should identify those features that science aims for and that bring about this success. As we have seen throughout this book, certainty is not such a feature. We have seen that we lack epistemic certainty when it comes to even the most mundane aspects of everyday life. If certainty were our goal in science, or in any other area of our lives, we would be doomed to fail. No current scientific theory is certain. What is more, we have seen that improved technology and better theories are not going to get us to certainty either. Certainty is unattainable.

To better understand why certainty is not an aim of science, let us briefly consider one of the most prominent debates in the philosophy of science, that between realists and anti-realists. Realists think that our current best scientific theories are true, or at least approximately true (this means that while they do not have all of the details correct, they are nevertheless getting the big picture right). Anti-realists deny this. They think that we do not have sufficient evidence to think that our current best scientific theories are actually true or even approximately true. At best, according to anti-realists, we can say that our current best theories are (mostly) correct in describing some observable phenomena. A central argument in this debate is what is known as the *pessimistic induction*.[14] It is an induction because it looks at the scientific theories that were the best in the past and generalizes by concluding that what was true of those theories is true of all of our best theories, including our current scientific theories. It is pessimistic because its conclusion is that our current best scientific theories (things like general relativity and evolution) are probably false, so we clearly should not think that they are true. The pessimistic induction supports this conclusion by noting that throughout the history of science all of our best scientific theories have turned out to be false. Since all of our previous best scientific theories have been false, we should think that our current best scientific theories are probably false, too.

There are a variety of realist replies to the pessimistic induction, and some are pretty persuasive.[15] However, for our purposes, it is enough to realize that even if anti-realists are incorrect and the pessimistic induction does not show that our current best scientific theories are likely to be false, it does provide yet another reason to accept that they are not certain. After all, no realist wants to say that the history of science provides no evidence whatsoever concerning the likelihood of our current best scientific theories being true or false. Clearly our past (now abandoned) best scientific theories were not certain; they were false! Likewise, our current best scientific theories are not certain. This is why (and this is the key point for our discussion) certainty cannot account for our scientific advancements. We are no more successful now at achieving certainty when it comes to any of our scientific theories than at any point in the history of science—we did not achieve certainty then, we still have not, and we never will.

Perhaps one might be tempted to think that while certainty is not a proper measure of scientific success, it is still an aim of science—the thought here being that science is successful as long as it produces theories that are closer to certain than those that came before. So, since our current best theories are closer to being certain than those that have come before, and the same is true of their predecessors and so on, science has been successful because, throughout its history, it has been getting closer to certainty. Admittedly, there is something correct about this idea. Our current best scientific theories are better supported by larger bodies of evidence than were their predecessors. They explain more data, make more accurate predictions, and so on. For example, general relativity explains the data that Newton's theory did, as well as accounting for phenomena that conflict with the predictions of Newtonian mechanics. One of the most famous of these is the perihelion precession of Mercury (in its orbit around the sun, Mercury travels in a way that conflicts with how Newtonian mechanics says it should move). Another is the gravitational redshift of light (gravity affects the frequency of light waves). As physicist Russell Stannard wrote, "General relativity accounts for *all* gravitational effects, including those that can be approximated by Newton's theory."[16] Accumulating more evidence and having stronger reasons to believe that our scientific theories are correct is surely a kind of success. But it is important to note that this is independent of whether we are aiming for certainty.

The success of science is better accounted for if we consider that the goal of science is deeper understanding of nature instead of certainty. Science thus succeeds when it produces scientific understanding. We can then easily see that science has continually advanced throughout its history because scientific understanding has increased, even if the explanations and predictions produced are uncertain. It can also occur even if the theories and models used to make those explanations and predictions are uncertain. In fact, such advancement can occur even if the theories and models involved are false! Elgin has nicely illustrated this point by noting how we have made advancements in astronomy despite using false theories.[17] Copernicus's theory improved upon Ptolemy's theory even though it included the idea that Earth's orbit around the sun is circular. Kepler's theory improved upon Copernicus's theory by holding that Earth's orbit was elliptical rather than circular. And Newton's theory of gravitation led him to make a further improvement by recognizing that Earth's orbit is not exactly elliptical. Each of these theories constituted a scientific advance from their predecessors. However, they were all false theories. Despite the fact that the theories were false—and thus clearly uncertain—they led to deeper understanding of Earth's orbit.

It is worth noting a key point here. Science often makes use of theories and models that are false, and yet they can yield genuine success in science.[18] The reason for this is that some false theories and models yield more understanding than other false theories and models. For example, a child who thinks that humans descended from apes, rather than that humans and apes descended from a common ancestor, is mistaken. Nevertheless, she has a deeper understanding of human evolution than a child who thinks that humans descended from butterflies.[19] Similarly, as we have mentioned in earlier chapters, science relies heavily on idealizations. Understanding can be facilitated by way of these idealizations. We can, and do, readily make use of models that are simplifications or distortions in order to better understand features of various phenomena. For instance, the Ideal Gas Law is a law describing the behavior of gases that do not exist. Despite this, the Ideal Gas Law provides understanding of how the pressure, volume, and temperature of actual gases interact in many cases. Idealizations and the uncertainty they bring with them are not an obstacle to understanding.

What is the upshot of all of this? Understanding, not certainty, is the aim of science. Scientists and philosophers think this for good reasons. Aiming at understanding is aiming at something that is more valuable than mere knowledge. Recognizing that science is trying to achieve deeper understanding of natural phenomena rather than certainty helps us to appreciate the tremendous success of science.

15 | How Uncertainty Makes Science Advance

> The very foundation of science is to keep the door open to doubt. Precisely because we keep questioning everything, especially our own premises, we are always ready to improve our knowledge. Therefore a good scientist is never "certain."
>
> —*Carlo Rovelli*[1]

Science Is Uncertain

Uncertainty is a feature of science in general for two main reasons. The first is simply that the world is extremely complex. When we are doing science, we are trying to understand complex phenomena such as obesity. There are numerous causal influences that affect whether or not someone is likely to be obese, and this is true of natural phenomena in general. The second fact leading to continual uncertainty in science is that humans are limited in important ways. We have limited access to relevant evidence. We cannot directly observe the past, and we cannot directly observe the future. We can only observe what is happening now, and, in fact, we can only observe a small portion of current happenings. Of course, the fact that we cannot directly observe the past or future does not mean that we cannot sufficiently explain the past or make accurate predictions about the future. It does, however, mean that our knowledge of past and future events is based on the evidence that we have access to in the present. Furthermore, as the *Uncertainty Principle* reminds us, there are some pieces of evidence that we simply cannot have, such as evidence concerning the exact location and momentum of an electron at a particular time. In addition to having limited access to evidence, we are limited in our ability to process the evidence that we do have. We are forced to rely on models to help us grasp the systems and phenomena that we seek to understand. And models always involve simplifying assumptions and idealizations, which generate uncertainties. As physicist Carlo Rovelli put it, "the core of science is not certainty, it's continual uncertainty."[2]

Although all science is uncertain, it does not follow that we should not trust scientific findings or what scientists tell us. Nor does it follow that all scientific theories are equally worthy of acceptance. Appreciating the uncertainties involved in science in general as well as particular scientific domains can help us to better understand what scientists are telling us and make better decisions on the basis of that information with respect to what to accept and how to act. Rather than a weakness of science, uncertainty and awareness of that uncertainty is a driving force for scientific advancement. As physicist Lawrence M. Krauss put it: "science has taught us over the years that certainty is largely an illusion. In science . . . [s]omething is either likely or unlikely, and we quantify *how* likely or unlikely. That is perhaps the greatest gift that science can give."[3] We cannot be certain about anything; but being aware of this and being able to estimate the level of our uncertainty allows us to successfully deal with it.

How Is Certainty Related to Rationality?

Another point that we have made throughout this book is that we do not need to have certainty in order to have knowledge, nor need we be certain in order to be reasonable in believing scientific theories and explanations or accepting predictions generated by those theories. It is worth taking some time here to look more closely at the relationship between certainty and rationality. Recall that there are two kinds of certainty: epistemic and psychological. *Epistemic certainty* is a matter of the strength of evidence that you have. You are epistemically certain of something when your evidence is so strong that it makes it *impossible* that you could have that evidence for the claim and be wrong. *Psychological certainty* is not a matter of the evidence one has. Instead, it concerns how strongly you believe something or how convinced you are that you are correct. Unfortunately, in many cases, people are psychologically certain even though their conviction is unrelated to the strength of their evidence.

How does this connect to rationality? Rationality amounts to proportioning one's beliefs to the evidence at hand. In other words, the rational attitude, when it comes to believing or accepting scientific theories (or anything else for that matter), is to have a confidence that matches the evidence in support of the theory. For instance, a theory like general relativity is extremely successful and is supported by a large body of evidence. As a result, we should be very confident in its veracity. Whereas if someone has recently come up with an untested hypothesis, we should not have much confidence that it is correct. The difference between this untested hypothesis and general relativity is simply a matter of evidence. In essence, rationality amounts to matching one's psychological conviction to one's epistemic situation. The closer one is to being epistemically certain about a theory, the closer one should be to psychological certainty about the theory. Perhaps this is better framed in terms of uncertainty, however, because we cannot have epistemic certainty. Put this way, rationality is a matter of matching our uncertainties to each other: our psychological uncertainty should match our epistemic uncertainty in the sense that the more we have of the latter, the more we should have of the former.

One might worry that this view of rationality will require that if we are to be rational, we will not be psychologically certain of much of anything. This is true, but it is not a cause for concern. As we see later in this chapter, remaining psychologically uncertain is not only the appropriate response to the evidence we have (i.e., our lack of epistemic certainty), but it is also a means to helping us continually improve our epistemic situation and advance science. Furthermore, owning up to our epistemic uncertainties by remaining psychologically uncertain to the appropriate degree is helpful. As Krauss has explained, "incorporating uncertainties prepares us to make more informed decisions about the future."[4] And more informed decisions of this sort are more rational decisions.

Dangers of Misunderstanding Scientific Uncertainty

In previous chapters, we briefly touched on ways that misunderstanding scientific uncertainty is problematic. It is time to delve more deeply into the problems caused by this sort of misunderstanding. In particular let us take a look at four significant dangers that we face when we misunderstand scientific uncertainty.

Danger 1: Dogmatic certainty. Obviously, epistemic certainty does not pose a danger. After all, there is no problem with having evidence that is so strong that there is no way you can be mistaken. However, as we have emphasized numerous times, epistemic certainty is not something that we can have in science (or likely anywhere else). Hence, the danger here refers to psychological certainty that is not matched by epistemic certainty—in other words, dogmatically accepting something as true. There are various negative consequences of holding scientific claims with psychological certainty. Let us consider three of them.

First of all, dogmatic certainty destroys rationality. Science does not admit of epistemic certainty; in other words, when it comes to the claims of science, it is always possible that we are mistaken. So there is always more evidence that could be had. As we saw in the previous section, rationality amounts to matching our level of psychological uncertainty to our level of epistemic uncertainty. If we are never epistemically certain of anything in science, then we should not be psychologically certain of anything in science either. After all, if we are psychologically certain of scientific claims, we are not matching our psychological conviction to the evidence we have. In other words, we are being irrational. This threat of irrationality always accompanies psychological certainty in science.

A second problem is that psychological certainty stops inquiry. If you are psychologically certain that a particular scientific claim is true, then you are convinced that any evidence that seems to go against that claim must be wrong or misleading. Why bother to look for more evidence when you are certain that the evidence will either support the claim or be misleading? This does not sound very scientific. Imagine a scientist who says that there is no evidence that could be found to make him give up a particular theory. That would just be a form of dogmatic fundamentalism. And such fundamentalism has no place in science. As Rovelli's claim in the epigraph made clear "a good scientist is never 'certain.'"[5]

The third problem that looms large when we are psychologically certain of scientific claims is intolerance. As neurologist Robert A. Burton has said, "it is in the leap from 99.99999 percent likely to a 100 percent guarantee that we give up tolerance for conflicting opinions, and provide the basis for the fundamentalist's claim to pure and certain knowledge."[6] After all, why should we consider opposing views when we are absolutely certain that we are correct? If there is really no possibility that we are wrong, then any contrary position must be mistaken. There would be no need to even consider the evidence for any other position. Again, this sort of intellectual intolerance amounts to dogmatic fundamentalism. Good science is not dogmatic, and good scientists (and laypeople) are "never 'certain'" about the claims of science.

Danger 2: Anything goes. On the other extreme from the previous danger, there is the danger of thinking that since science is uncertain, anything goes. One might be tempted to think that since scientists are not certain that any particular theory is absolutely correct, all theories are in the same boat. This, of course, is a serious mistake. Thinking that anything goes treats evidence in an all-or-nothing manner: either the evidence makes a scientific claim certain or it does not support that claim over any other. Clearly, this is misguided. Consider, when it looks like it is raining outside, the weather person on television says that it is raining, and it sounds like it is raining outside, it still is not epistemically certain that it is raining. It is at least *possible* that despite all of this evidence it is not raining. After all, it could look and sound like it is raining because someone is spraying down the roof for some reason (e.g., to clean it). Is it likely that by coincidence unknown to you both someone is spraying down the roof and the weather person misspeaks about the current weather conditions? No, but it is *possible*. And, insofar as it is possible, the evidence you have does not make it epistemically certain that it is raining. Nevertheless, the evidence in this sort of case supports that it is in fact raining. Given this evidence (and no reason to think someone is spraying the roof or that the weather person misspoke), it would be rational to think that it is raining.

We can see why thinking that any scientific claim is as good as another is particularly harmful by considering a recently proposed alternative explanation for climate change. As we have mentioned in earlier chapters, the prevailing view is that climate change is largely the result of human emissions of greenhouse gases such as CO_2. Some have attempted to undermine support for this explanation of rising average global temperatures by proposing that increased solar activity is instead responsible for the temperature increases. If we were to take the fact that scientific claims are never certain to mean that anything goes, we might be inclined to think that increased solar activity is just as good of an explanation of increased temperatures as human greenhouse gas emissions. If the two are equally good explanations of rising temperatures, then there is no good reason to think that one rather than the other is correct. If one thinks that anything goes, then it seems that we are equally well off accepting either proposed explanation. This is a very serious mistake though. We saw in Chapter 6 that the evidence strongly suggests both that greenhouse gas emissions have sharply increased along with average global temperatures, and there has been no similar increase in solar activity in recent years.[7] Consequently, these two explanations are not equally good. The evidence supports the claim that humans are causing climate change; it does not support the claim that climate change is the result of increased solar activity. Uncertainty in science does not mean that anything goes. It means that we should always be open to new evidence and revision, but that, in the meantime, we should judge scientific claims on the basis of the evidence available. And not all claims are equally well supported by the evidence. Well-supported claims are the only ones we should accept, even though they are not certain.

Danger 3: Misunderstanding sciences. It is worth emphasizing something that we saw in Chapters 6–10: when we misunderstand the uncertainty involved in particular sciences, we misunderstand and are prone to misjudge those sciences. In some cases misunderstanding the role of uncertainty may lead one to dismiss the underlying science. This is something that we see with climate change and human evolution. In other cases, uncertainty may lead one to forego very beneficial and evidence-based medical interventions, as we have seen in the case of vaccination. At the other extreme, we might misunderstand the sciences by expecting more certainty from them than they can provide. We see this sort of problem particularly well when it comes to genetic testing and forensic science. In each case misunderstanding the uncertainties involved leads to misunderstanding the science and what the methods of that science can tell us. Anything that we can know comes with a degree of uncertainty. In all the case studies discussed earlier, and in all sciences, we can be quite confident of particular conclusions. We should neither reject these conclusions because uncertainties exist nor reach for further conclusions beyond what the evidence supports.

Danger 4: Deceit. A final danger to discuss here is one that can occur anytime there is misunderstanding: deceit. When there is something that we fail to understand, others who do understand it can mislead us if they choose. Unfortunately, this is something that has happened in connection with things that pose serious health risks. As historians of science Naomi Oreskes and Erik M. Conway have convincingly argued in their book, *Merchants of Doubt*, a group of scientists and science advisors, for political and/or financial reasons, misled the public about both the dangers of tobacco smoke and the threat of climate change and humans' role in it.[8] How could the public be misled about either of these, given the vast amounts of evidence available showing that tobacco smoke is a cause of cancer and other illnesses and that humans are causing climate change? A prominent factor in this was (and, unfortunately, still is) misunderstanding the role of uncertainty in science. Since many people do not understand that all science is uncertain, pointing out uncertainties in specific ways can mislead people into thinking that the links between things like smoking and cancer or greenhouse gases and climate change are not well-established. The way to safeguard ourselves against such trickery is to appreciate and accept that uncertainty is inherent in science. The question therefore should not be whether uncertainties exist but whether they affect our understanding of the big picture. If such understanding is achieved, we can live with uncertainties about the details.[9]

Uncertainty Makes Science Advance

Throughout this book we have argued that science is uncertain but also that this is not a cause for concern. Counterintuitive though it might at first seem, uncertainty in science is an advantage. We have explained in several specific instances how uncertainty helps science advance. It is worth taking a bit of time here at the end to draw some general conclusions about the role that uncertainty plays in the advancement of science. There are several ways that uncertainty (especially when recognized and appreciated) is beneficial to science and our understanding of science. The first way is that it helps one to avoid the dangers discussed in the previous section. Someone who appreciates the uncertainty that is always present in science is less apt to fall into the trap of accepting scientific claims with dogmatic certainty. Additionally, such a person will be inclined to appreciate the role that evidence plays in determining what we should accept. As a result, one will be less likely to assume that anything goes when it comes to science. Furthermore, by understanding the uncertainties of particular sciences, one goes a long way toward understanding those sciences. Finally, with a proper appreciation of uncertainty in science, one is much harder to deceive about the real standing of various scientific claims.

In addition to helping us avoid the dangers just discussed, appreciating uncertainty provides a better understanding of our current epistemic situation.[10] When we take account of uncertainties in the present, we can better grasp what it is that we should be confident of and what we should not, and we are in a position to make more informed decisions about the future. For example, if we know the uncertainties involved in having our child receive a particular common vaccine—it has a very high probability of preventing a life-threatening illness and a very low probability of having any harmful side effects—we can weigh the odds of the potential benefits versus the odds of the potential costs. Doing this allows us to make a rational decision that will be apt to provide the best outcome for our child. The result of being aware of the uncertainties is that we are in a position to appreciate the information we have and make the best decision possible despite lacking certainty.

Uncertainty in science not only provides insight into our current epistemic situation, it also actually helps us to achieve the central epistemic aim of science: understanding. Since it is uncertain, science is continually searching for new evidence. If we are certain, we stop searching for further evidence. Such evidence can provide us with even better support for the theories we accept, or it can lead us to new theories and discoveries. Regardless of whether the new evidence supports current theories or challenges us to change our minds, the result is a deeper understanding of the universe. And it is uncertainty that drives this continual search for evidence.

There are undoubtedly additional ways that uncertainty helps science advance, especially particular uncertainties that arise in the context of specific scientific inquiries. However, we have seen the big picture of how uncertainty makes science advance. Hopefully, this is enough to help us look to uncertainty not with trepidation but with appreciation. In light of the fact that his two books,

Failure and *Ignorance*,[11] were our inspiration for writing this book, it is fitting that we close with a quote from neurobiologist Stuart Firestein that sums up how we should approach revelations of uncertainty in science:

> Expressing doubt and uncertainty should make a person more trustworthy. Anyone who claims to know the truth, the Truth, because they are special or have a special connection to some authority no one can question, these are the people you want to be wary of. . . . Science is the best method I know for being wary without being paranoid.[12]

NOTES

Chapter 1

1. Smyth, Albert Henry, *The Writings of Benjamin Franklin, Volume X, 1789–1790, Collected and Edited with a Life and Introduction.* London: Macmillan, 1907, p. 69. https://archive.org/details/writingsofbenjam10franuoft. Accessed April 28, 2018.

2. Things are actually a bit more complicated. Most epistemologists now think that knowledge requires something more than a true belief with sufficiently strong evidence, though they do not agree on what exactly this additional factor is. For our purposes, we do not need to delve into these more esoteric points of epistemology (see McCain, Kevin, *The Nature of Scientific Knowledge: An Explanatory Approach.* Dordrecht: Springer, 2016, chapter 8).

3. In this case, some of this evidence would take the form of satellite photos, for instance, https://www.nasa.gov/image-feature/nasa-captures-epic-earth-image. Accessed April 26, 2018.

4. https://theflatearthsociety.org/home/. Accessed April 26, 2018.

5. https://theflatearthsociety.org/home/index.php/blog/lunar-eclipses-and-shadow-earth. Accessed April 26, 2018.

6. https://www.youtube.com/watch?v=hkdgaEKO-SQ. Accessed April 26, 2018.

7. Descartes, René, "Meditations on First Philosophy," in *Descartes: Selected Philosophical Writings*, J. Cottingham, R. Stoothoff, and D. Murdoch (Trans.). Cambridge: Cambridge University Press, 1641/1988, pp. 73–123.

8. https://www.imdb.com/title/tt0120382/. Accessed April 26, 2018.

9. Kahneman, Daniel, *Thinking, Fast and Slow*. New York: Farrar, Straus and Giroux, 2011, pp. 156–158.

10. Wason, Peter, "Reasoning," in *New Horizons in Psychology*, Brian M. Foss (ed.). New York: Penguin, 1966, pp. 135–151.

11. Stein, Edward, *Without Good Reason: The Rationality Debate in Philosophy and Cognitive Science*. Oxford: Clarendon Press, 2001, p. 82.

12. Gilovich, Thomas, *How We Know What Isn't So: The Fallibility of Human Reason in Everyday Life*. New York: Free Press, 1991, p. 77.

13. Mercier, Dan, and Hugo Sperber, *The Enigma of Reason*. Cambridge, MA: Harvard University Press, 2017, p. 218.

14. Hertwig, Ralph, and Gerd Gigerenzer, "The 'Conjunction Fallacy' Revisited: How Intelligent Inferences Look Like Reasoning Errors," *Journal of Behavioral Decision Making*, 12 (1999): 275–305.

15. Mousavi, Shabnam, and Gerd Gigerenzer, "Revisiting the "Error" in Studies of Cogelnitive Errors," in *Error in Organizations*, D. A. Hofmann and M. Frese (eds.). New York: Taylor and Francis, 2011, pp. 97–112.

16. Kahneman, *Thinking, Fast and Slow*, p. 4.

17. Gigerenzer, Gerd, "On the Supposed Evidence for Libertarian Paternalism," *Review of Philosophy and Psychology*, 6 (2015): 361–383.

18. Isidore, Chris, "These Are Your Odds of Winning Powerball or Mega Millions," CNNMoney. http://money.cnn.com/2018/01/04/news/powerball-mega-millions-odds/index.html. Accessed April 25, 2018.

Chapter 2

1. Burton, Robert Alan, *On Being Certain: Believing You Are Right Even When You're Not*. New York: St. Martins Griffin, 2008, p. xiii.

2. Principe, Lawrence M., "Myth 4. That Alchemy and Astrology Were Superstitious Pursuits That Did Not Contribute to Science and Scientific Understanding," in *Newton's Apple and Other Myths About Science*, Ronald L. Numbers and Kostas Kampourakis (eds.). Cambridge MA: Harvard University Press, 2015, pp. 32–36.

3. Lyons, Linda, "Paranormal Beliefs Come (Super)Naturally to Some," Gallup. Washington, DC. November 1, 2005. http://www.gallup.com/

poll/19558/paranormal-beliefs-come-supernaturally-some.aspx. Accessed May 23, 2018.

4. Matthew 7:12 (NIV).

5. Proverbs 27:1 (NIV).

6. James 4:14 (NIV).

7. Houk, James T., *The Illusion of Certainty: How the Flawed Beliefs of Religion Harm Our Culture*. Amherst, NY: Prometheus Books, 2017, p. 12.

8. Burton, *On Being Certain*, pp. 8–9.

9. Ibid., pp. 12–15.

10. Ibid., p. 40

11. Ibid., p. xiv

12. Seidel, Eva-Maria, Daniela M. Pfabigan, Andreas Hahn, Ronald Sladky, Arvina Grahl, Katharina Paul, Christoph Kraus, et al., "Uncertainty During Pain Anticipation: The Adaptive Value of Preparatory Processes," *Human Brain Mapping* 36, no. 2 (2015): 744–755.

13. De Berker, Archy O., Robb B. Rutledge, Christoph Mathys, Louise Marshall, Gemma F. Cross, Raymond J. Dolan, and Sven Bestmann, "Computations of Uncertainty Mediate Acute Stress Responses in Humans," *Nature Communications* 7 (2016): 10996.

14. Hsu, Ming, Meghana Bhatt, Ralph Adolphs, Daniel Tranel, and Colin F. Camerer, "Neural Systems Responding to Degrees of Uncertainty in Human Decision-Making," *Science* 310, no. 5754 (2005): 1680–1683.

15. Gigerenzer, Gerd, *Risk Savvy: How To Make Good Decisions*. New York: Viking, 2014, p. 14.

16. Burton, *On Being Certain*, p. 99.

17. American Cancer Society, *Cancer Facts and Figures 2016*. Atlanta, GA: American Cancer Society, 2016, p. 11.

18. http://www.dailymail.co.uk/health/article-3052396/Women-like-Angelina-Jolie-carry-BRCA1-gene-likely-die-breast-cancer-OVARIES-removed.html. Accessed May 28, 2018.

19. https://edition.cnn.com/2014/02/25/health/brca-study/index.html. Accessed May 28, 2018.

20. http://www.foxnews.com/health/2015/10/18/moms-with-brca-breast-cancer-gene-mutations-face-tough-decisions.html. Accessed May 28, 2018.

Chapter 3

1. https://twitter.com/realDonaldTrump/status/338978381636984832, May 27, 2013. Accessed March 28, 2018.

2. https://www.vox.com/policy-and-politics/2017/6/1/15726472/trump-tweets-global-warming-paris-climate-agreement. Accessed March 28, 2018.

3. https://twitter.com/realDonaldTrump/status/264441602636906496. Accessed March 28, 2018.

4. IPCC, *Climate Change 2014: Synthesis Report. Contribution of Working Groups I, II and III to the Fifth Assessment Report of the Intergovernmental Panel on Climate Change.* Core Writing Team, R. K. Pachauri and L. A. Meyer (eds.) Geneva, Switzerland: IPCC, 2014, 151 pages. (The numbers in the text come from the Summary for Policy Makers.)

5. https://www.nytimes.com/2017/04/28/opinion/climate-of-complete-certainty.html. Accessed March 29, 2018.

6. Pew Research Center, October, 2016, "The Politics of Climate"; http://www.pewinternet.org/2016/10/04/the-politics-of-climate/. Accessed March 29, 2018.

7. Scientism is, simply put, the attribution of extreme authority to science and the attitude of privileging science over any other way of knowing (see Boudry, Maarten, and Massimo Pigliucci, eds., *Science Unlimited?: The Challenges of Scientism*. Chicago: University of Chicago Press, 2018).

8. https://www.climatefactsfirst.org. Accessed March 29, 2018.

9. Zehr, Stephen C., "Public Representations of Scientific Uncertainty About Global Climate Change," *Public Understanding of Science* 9, no. 2 (2000): 85–103.

10. Retzbach, Andrea, and Michaela Maier, "Communicating Scientific Uncertainty: Media Effects on Public Engagement with Science," *Communication Research* 42, no. 3 (2015): 429–456.

11. Broomell, Stephen B., and Patrick B. Kane, "Public Perception and Communication of Scientific Uncertainty," *Journal of Experimental Psychology: General* 146, no. 2 (2017): 286.

12. Retzbach, Joachim, Andrea Retzbach, Michaela Maier, Lukas Otto, and Marion Rahnke, M., "Effects of Repeated Exposure to Science TV Shows on Beliefs About Scientific Evidence and Interest in Science," *Journal of Media Psychology* 25, no. 1 (2013): 3–13.

13. For an excellent and detailed account, see Oreskes, Naomi, and Erik M. Conway, *Merchants of Doubt: How a Handful of Scientists Obscured the Truth on Issues from Tobacco Smoke to Global Warming.* New York: Bloomsbury, 2011.

14. Lewandowsky, Stephan, Gilles E. Gignac, and Samuel Vaughan, "The Pivotal Role of Perceived Scientific Consensus in Acceptance of Science," *Nature Climate Change* 3, no. 4 (2013): 399–404.

15. Salmon, Daniel A., Matthew Z. Dudley, Jason M. Glanz, and Saad B. Omer, "Vaccine Hesitancy: Causes, Consequences, and a Call to Action," *Vaccine* 33 (2015): D66–D71.

16. Wood, Bernard, and Eve K. Boyle, "Hominin Taxic Diversity: Fact or Fantasy?" *American Journal of Physical Anthropology* 159, no. S61 (2016): 37–78; Wood, Bernard, "Evolution: Origin(s) of Modern Humans," *Current Biology* 27, no. 15 (2017): R767–769.

17. Lander, Eric S., "Cutting the Gordian Helix: Regulating Genomic Testing in the Era of Precision Medicine," *New England Journal of Medicine* 372 (2015): 1185–1186.

18. Gigerenzer, Gerd, *Risk Savvy: How to Make Good Decisions.* New York: Viking, 2014, pp. 23–32.

19. Gigerenzer, Gerd, Ralph Hertwig, Eva Van Den Broek, Barbara Fasolo, and Konstantinos V. Katsikopoulos, "'A 30% Chance of Rain Tomorrow': How Does the Public Understand Probabilistic Weather Forecasts?" *Risk Analysis* 25, no. 3 (2005): 623–629.

20. Gigerenzer, *Risk Savvy*, pp. 3–5.

Chapter 4

1. https://www.youtube.com/watch?timecontinue=8andv=MmpUWEW6Is. Accessed April 28, 2018.

2. Nobelprize.org, "Richard P. Feynman—Nominations," Nobel Media AB. 2014. http://www.nobelprize.org/nobelprizes/physics/laureates/1965/feynman-nomination.html. Accessed April 28, 2018.

3. Nobelprize.org, "The Nobel Prize in Physics 1965," Nobel Media AB. 2014. http://www.nobelprize.org/nobelprizes/physics/laureates/1965/. Accessed April 28, 2018.

4. http://www.richardfeynman.com/works/popular.html. Accessed April 28, 2018.

5. https://www.youtube.com/watch?v=6Rwcbsn19c0. Accessed April 28, 2018.

6. We follow the definition and analysis of scientific expertise found in Nichols, Thomas M., *The Death of Expertise: The Campaign against Established Knowledge and Why It Matters*. New York: Oxford University Press, 2017, pp. 28–39.

7. Stathis, Psillos, "Is the History of Science the Wasteland of False Theories?" in *Adapting Historical Science Knowledge Production to the Classroom*, P. V. Kokkotas et al. (eds.). Rotterdam, Netherlands: Sense Publishers, 2011, pp. 17–36 at p. 32.

8. Smith, Tom W., and Jaesok Son, *Trends in Public Attitudes about Confidence in Institutions. General Social Survey 2012 Final Report.* Chicago: NORC at the University of Chicago, 2013, p.14.

9. Ibid., pp. 6–8.

10. Pew Research Center. October, 2016, "The Politics of Climate," http://assets. pewresearch.org/wp-content/uploads/sites/14/2016/10/14080900/ PS2016.10.04Politics-of-ClimateTOPLINE.pdf. Accessed May 3, 2017.

11. Ibid.

12. Ibid., pp. 5–7, 35.

13. Ibid., pp. 68–73.

14. Ibid., pp. 73–75.

15. http://news.gallup.com/poll/231530/global-warming-concern-steady-despite-partisan-shifts.aspx. Accessed May 7, 2018.

Chapter 5

1. Ioannidis, John P. A., "Why Most Published Research Findings Are False," *PLoS Medicine* 2, no. 8 (2005): e124, 0700.

2. https://profiles.stanford.edu/john-ioannidis. Accessed May 13, 2018.

3. https://scholar.google.com/citations?user=A9e6sPYAAAAJandhl=de. Accessed December 19, 2018.

4. https://profiles.stanford.edu/john-ioannidis. Accessed May 13, 2018.

5. Ioannidis, John P. A., "Why Most Discovered True Associations Are Inflated," *Epidemiology* 19 (2008): 640–648.

6. Contopoulos-Ioannidis, Despina G., George A. Alexiou, Theodore C. Gouvias, and John P. Ioannidis, "Life Cycle of Translational Research for Medical Interventions," *Science* 321 (2008): 1298–1299.

7. Ioannidis, John P. A., "Why Most Clinical Research Is Not Useful," *PLoS Medicine* 13, no. 6 (2016): e1002049.

8. Ioannidis, John P. A., "All Science Should Inform Policy and Regulation," *PLoS Medicine* 15, no. 5 (2018): e1002576, p. 1

9. https://retractionwatch.com. Accessed May 17, 2018.

10. Harris, Richard, *Rigor Mortis: How Sloppy Science Creates Worthless Cures, Crushes Hope, and Wastes Billions*. New York: Basic Books, 2017, p. 181.

11. Fanelli, Daniele, "How Many Scientists Fabricate and Falsify Research? A Systematic Review and Meta-Analysis of Survey Data," *PLoS One* 4, no. 5 (2009): e5738.

12. Silberzahn, Raphael, and Eric L. Uhlmann, "Many Hands Make Tight Work," *Nature* 526, no. 7572 (2015): 189.

13. Begley, C. Glenn, and Lee M. Ellis, "Drug Development: Raise Standards for Preclinical Cancer Research," *Nature* 483, no. 7391 (2012): 531.

14. Open Science Collaboration, "Estimating the Reproducibility of Psychological Science," *Science* 349, no. 6251 (2015): aac4716.

15. Baker, Monya, "Is There a Reproducibility Crisis?" *Nature* 533 (2016): 452–454.

16. Goodman, Steven N., Daniele Fanelli, and John P. A. Ioannidis, "What Does Research Reproducibility Mean?" *Science Translational Medicine* 8, no. 341 (2016): 341ps12–341ps12.

17. Yamada, Kenneth M., and Alan Hall, "Reproducibility and Cell Biology," *The Journal of Cell Biology* 209, no. 2 (2015): 191–193.

18. For a specific example in cell biology, see Madsen, Daniel H., and Thomas H. Bugge, "The Source of Matrix-Degrading Enzymes in Human

Cancer: Problems of Research Reproducibility and Possible Solutions," *Journal of Cell Biology* 209, no. 2 (2015): 195–198.

19. https://www.nature.com/articles/d41586-018-02563-4;   http:// science.sciencemag.org/content/360/6385/133.full.   Both   accessed May 17, 2018.

20. Ioannidis, John P. A., Richard Klavans, and Kevin W. Boyack, "The Scientists Who Publish a Paper Every Five Days," *Nature* 561 (2018): 167–169.

21. Fanelli, Daniele, and Vincent Larivière, "Researchers' Individual Publication Rate Has Not Increased in a Century," *PLoS One* 11 (2016): e0149504.

22. https://www.youtube.com/watch?v=0Rnq1NpHdmw.   Accessed   May 17, 2018.

23. Fanelli, Daniele, "Do Pressures to Publish Increase Scientists' Bias? An Empirical Support from US States Data," *PloS One* 5, no. 4 (2010): e10271.

24. Fanelli, Daniele, "Why Growing Retractions Are (Mostly) a Good Sign," *PLoS Med* 10 (2013): e1001563.

25. Redish, A. David, Erich Kummerfeld, Rebecca Lea Morris, and Alan C. Love, "Opinion: Reproducibility Failures Are Essential to Scientific Inquiry," *Proceedings of the National Academy of Sciences* 115, no. 20 (2018): 5042–5046.

Chapter 6

1. Maslin, Mark, *Climate Change: A Very Short Introduction*. Oxford: Oxford University Press, 2014, p. 56.

2. US President Barack Obama, interview with Al Roker, May 6, 2014. https://www.today.com/news/obama-al-roker-climate-change-problem-affecting-americans-right-now-2D79627268. Accessed July 12, 2018.

3. NASA, "What's in a Name? Global Warming vs. Climate Change." 2008. https://pmm.nasa.gov/education/articles/whats-name-global-warming-vs-climate-change. Accessed July 12, 2018.

4. https://twitter.com/realDonaldTrump/status/316252016190054400. Accessed July 12, 2018.

5. IPCC, *Climate Change 2014: Synthesis Report. Contribution of Working Groups I, II and III to the Fifth Assessment Report of the Intergovernmental Panel on Climate Change*. Core Writing Team, R. K. Pachauri and L. A.

Meyer (eds.) Geneva, Switzerland: IPCC, 2014, p. 10. (The numbers in the text come from the Summary for Policy Makers.)

6. Harker, David, *Creating Scientific Controversies: Uncertainty and Bias in Science and Society*. Cambridge: Cambridge University Press, 2015, p. 180.

7. Romm, Joseph, *Climate Change: What Everyone Needs to Know,* 2nd edition. New York: Oxford University Press, 2018, p. 3.

8. Ibid., p. 2.

9. Ibid., p. 31.

10. IPCC. *Climate Change 2014,* p. 2.

11. Oreskes, Naomi, "The Scientific Consensus on Climate Change: How Do We Know We're Not Wrong?" in *Climate Change: What It Means to Us, Our Children, and Our Grandchildren*, Joseph F. C. Dimento and Pamela Doughman (eds.). Cambridge, MA: MIT Press, 2007, pp. 65–99 at p. 82 .

12. Stoller-Conrad, Jessica, "Tree Rings Provide Snapshots of Earth's Past Climate," NASA Global Climate Change. January 25, 2017. https://climate.nasa.gov/news/2540/tree-rings-provide-snapshots-of-earths-past-climate/. Accessed August 15, 2018.

13. Stoller-Conrad, Jessica, "Core Questions: An Introduction to Ice Cores," NASA Global Climate Change. August 14, 2017. https://climate.nasa.gov/news/2616/core-questions-an-introduction-to-ice-cores/. Accessed August 15, 2018.

14. IPCC. *Climate Change 2014,* p. 2.

15. Romm, *Climate Change*, p. 2.

16. IPCC. *Climate Change 2014,* p. 5.

17. Maslin. *Climate change: A Very Short Introduction,* pp. 7–9.

18. IPCC. *Climate Change 2014,* p. 2.

19. Romm, *Climate Change*, p. 3.

20. Ibid., p. 9.

21. IPCC. *Climate Change 2014,* p. 2.

22. Oreskes, Naomi, "The Scientific Consensus on Climate Change," *Science* 306, no. 1686 (2004): 1686.

23. Oreskes, "The Scientific Consensus on Climate Change: How Do We Know We're Not Wrong?" p. 74.

24. Potochnik, Angela, *Idealization and the Aims of Science*. Chicago: University of Chicago Press, 2017, p. 84.

25. Ibid., p. 205.

26. Ibid., p. 85.

27. Rahmstorf, Stefan, Grant Foster, and Anny Cazenave, "Comparing Climate Projections to Observations up to 2011," *Environmental Research Letters* 7, no. 044035 (2012). https://iopscience.iop.org/article/10.1088/1748-9326/7/4/044035/pdf. Accessed July 16, 2018.

28. Maslin, *Climate Change: A Very Short Introduction,* pp. 54–57.

29. Leiserowitz, Anthony, Edward Maibach, Connie Roser-Renouf, Seth Rosenthal, Matthew Cutler, and John Kotcher, *Climate Change in the American Mind: March 2018*. Yale University and George Mason University. New Haven, CT: Yale Program on Climate Change Communication, 2017. http://climatecommunication.yale.edu/publications/climate-change-american-mind-march-2018/3/. Accessed July 16, 2018.

30. Oreskes, "The Scientific Consensus on Climate Change", p. 1686.

31. Cook, John, Dana Nuccitelli, Sarah A. Green, Mark Richardson, Barbel Winkler, Rob Painting, et al., "Quantifying the Consensus on Anthropogenic Global Warming in the Scientific Literature," *Environmental Research Letters* 8, no. 024024 (2013). http://iopscience.iop.org/article/10.1088/1748-9326/8/2/024024. Accessed July 16, 2018.

32. Oreskes, "The Scientific Consensus on Climate Change," p. 1686.

33. Oreskes, "The Scientific Consensus on Climate Change: How Do We Know We're Not Wrong?" p. 74.

34. Oreskes, Naomi, and Erik M. Conway, *Merchants of Doubt: How a Handful of Scientists Obscured the Truth on Issues from Tobacco Smoke to Global Warming*. New York: Bloomsbury Press, 2010.

35. Oreskes, "The Scientific Consensus on Climate Change: How Do We Know We're Not Wrong?" p. 74.

36. Of course, these points are not restricted to climate science. The realities of any scientific issue can be obscured by those with political or financial incentives to do so or by poor media representation.

Chapter 7

1. https://www.pbs.org/wgbh/frontline/article/paul-offit-a-choice-not-to-get-a-vaccine-is-not-a-risk-free-choice/. Accessed August 31, 2018.

2. We are actually immune to pathogens, not to diseases, and it is pathogens, not diseases, that can be transmitted.

3. Orenstein, Walter A., and Rafi Ahmed, "Simply Put: Vaccination Saves Lives," *Proceedings of the National Academies of Science* 114, no. 16 (2017): 4031–4033.

4. Offit, Paul A., *Deadly Choices: How the Anti-Vaccine Movement Threatens Us All*. New York: Basic Books, 2015, pp. xxii–xxiii.

5. https://www.fda.gov/biologicsbloodvaccines/safetyavailability/vaccinesafety/ucm187810.htm. Accessed August 27, 2018; Offit, Paul A., and Rita K. Jew, "Addressing Parents' Concerns: Do Vaccines Contain Harmful Preservatives, Adjuvants, Additives, or Residuals?" *Pediatrics* 112, no. 6 (2003): 1394–1397.

6. Maglione, Margaret A., Lopamudra Das, Laura Raaen, Alexandria Smith, Ramya Chari, Sydne Newberry, Roberta Shanman, Tanja Perry, Matthew Bidwell Goetz, and Courtney Gidengil, "Safety of Vaccines Used for Routine Immunization of US Children: A Systematic Review," *Pediatrics* 134 (2014): 325–337.

7. Zhou, Fangjun, Abigail Shefer, Jay Wenger, Mark Messonnier, Li Yan Wang, Adriana Lopez, Matthew Moore, Trudy V. Murphy, Margaret Cortese, and Lance Rodewald, "Economic Evaluation of the Routine Childhood Immunization Program in the United States, 2009," *Pediatrics* 133 (2014): 577–585.

8. https://www.youtube.com/watch?v=qpUsg4bDH5w. Accessed September 11, 2018.

9. Offit, *Deadly Choices*, chapter 1.

10. Oshinsky, David M., *Polio: An American Story*. Oxford: Oxford University Press, 2005, p. 237.

11. Offit, Paul A., *The Cutter Incident: How America's First Polio Vaccine Led to the Growing Vaccine Crisis*. New Haven, CT: Yale University Press, 2007.

12. https://www.cdc.gov/pertussis/surv-reporting/cases-by-year.html. Accessed October 30, 2018.

13. Wakefield, Andrew J., Simon H. Murch, Andrew Anthony, John Linnell, D. M. Casson, Mohsin Malik, Mark Berelowitz, et al., "Retracted: Ileal-Lymphoid-Nodular Hyperplasia, Non-Specific Colitis, and Pervasive Developmental Disorder in Children," *The Lancet* 351, no. 9103 (1998): 637–641, at p. 638.

14. Ibid., at p. 641.

15. Chen, Robert T., and Frank DeStefano, "Vaccine Adverse Events: Causal or Coincidental?" *The Lancet* 351, no. 9103 (1998): 611–612.

16. Afzal, M. A., P. D. Minor, J. Begley, M. L. Bentley, E. Armitage, S. Ghosh, and A. Ferguson, "Absence of Measles-Virus Genome in Inflammatory Bowel Disease," *The Lancet* 351 (1998): 646–647.

17. See "Autism, Inflammatory Bowel Disease, and MMR Vaccine," *The Lancet* 351 (1998): 905–908.

18. Wakefield, Andrew J., "Author's Reply," *The Lancet* 351 (1998): 907–908.

19. https://briandeer.com/mmr/lancet-deer-1.htm; https://www.thetimes. co.uk/article/revealed-mmr-research-scandal-7ncfntn8mjq. Both accessed October 30, 2018.

20. For details, see Deer, Brian, "How the Case Against MMR Vaccine Was Fixed," *British Medical Journal* 342 (2011 Jan. 5): c5347; Deer, Brian, "How the Vaccine Crisis Was Meant to Make Money," *British Medical Journal* 342 (2011 Jan. 11): c5258; Deer, Brian, "The Lancet's Two Days to Bury Bad News," *British Medical Journal* 342 (2011 Jan 18): c7001.

21. https://briandeer.com/solved/gmc-wakefield-sentence.pdf.   Accessed November 9, 2018.

22. The editors of *The Lancet* "Retraction: Ileal-Lymphoid-Nodular Hyperplasia, Non-Specific Colitis, and Pervasive Developmental Disorder in Children," *The Lancet* 375 (2010): 445.

23. Goldacre, Ben, *Bad Science*. London: Fourth Estate, 2009, pp. 290–291.

24. Ibid., p. 306.

25. Afzal, M. A., L. C. Ozoemena, A. O'hare, K. A. Kidger, M. L. Bentley, and P. D. Minor, "Absence of Detectable Measles Virus Genome Sequence

in Blood of Autistic Children Who Have Had Their MMR Vaccination During the Routine Childhood Immunization Schedule of UK," *Journal of Medical Virology* 78, no. 5 (2006): 623–630.

26. Peltola, Heikki, Annamari Patja, Pauli Leinikki, Martti Valle, Irja Davidkin, and Mikko Paunio, "No Evidence for Measles, Mumps, and Rubella Vaccine-Associated Inflammatory Bowel Disease or Autism in a 14-Year Prospective Study," *The Lancet* 351, no. 9112 (1998): 1327–1328; Taylor, Brent, Elizabeth Miller, C. Paddy Farrington, Maria Christina Petropoulos, Isabelle Favot-Mayaud, Jun Li, et al., "Autism and Measles, Mumps, and Rubella Vaccine: No Epidemiologic Evidence for a Causal Association," *The Lancet* 353 (1999): 2026–2029; Smeeth, Liam, Claire Cook, Eric Fombonne, Lisa Heavey, Laura C. Rodrigues, Peter G. Smith, and Andrew J. Hall, "MMR Vaccination and Pervasive Developmental Disorders: A Case-Control Study," *The Lancet* 364, no. 9438 (2004): 963–969; Madsen, Kreesten M., Anders Hviid, Mogens Vestergaard, Diana Schendel, Jan Wohlfahrt, Poul Thorsen, Jørn Olsen, and Mads Melbye, "A Population-Based Study of Measles, Mumps, and Rubella Vaccination and Autism," *New England Journal of Medicine* 347, no. 19 (2002): 1477–1482; Mrozek-Budzyn, Dorota, Agnieszka Kieltyka, and Renata Majewska, "Lack of Association Between Measles-Mumps-Rubella Vaccination and Autism in Children: A Case-Control Study," *The Pediatric Infectious Disease Journal* 29, no. 5 (2010): 397–400; Taylor, Luke E., Amy L. Swerdfeger, and Guy D. Eslick, "Vaccines Are Not Associated with Autism: An Evidence-Based Meta-Analysis of Case-Control and Cohort Studies," *Vaccine* 32, no. 29 (2014): 3623–3629; Jain, Anjali, Jaclyn Marshall, Ami Buikema, Tim Bancroft, Jonathan P. Kelly, and Craig J. Newschaffer, "Autism Occurrence by MMR Vaccine Status Among US Children with Older Siblings with and without Autism," *Journal of the American Medical Association* 313, no. 15 (2015): 1534–1540; Gadad, Bharathi S., Wenhao Li, Umar Yazdani, Stephen Grady, Trevor Johnson, Jacob Hammond, Howard Gunn, et al., "Administration of Thimerosal-Containing Vaccines to Infant Rhesus Macaques Does Not Result in Autism-Like Behavior or Neuropathology," *Proceedings of the National Academy of Sciences* 112, no. 40 (2015): 12498–12503.

27. Glass, Roger I., and Umesh D. Parashar, "Rotavirus Vaccines: Balancing Intussusception Risks and Health Benefits," *The New England Journal of Medicine* 370, no. 6 (2014): 568.

28. https://vimeo.com/230115723. Accessed November 9, 2018.

29. Baay, Marc, Kaatje Bollaerts, and Thomas Verstraeten, "A Systematic Review and Meta-Analysis on the Safety of Newly Adjuvanted Vaccines Among Older Adults," *Vaccine* 36, no. 29 (2018): 4207–4214.

30. Flacco, Maria Elena, Lamberto Manzoli, Annalisa Rosso, Carolina Marzuillo, Mauro Bergamini, Armando Stefanati, Rosario Cultrera, et al., "Immunogenicity and Safety of the Multicomponent Meningococcal B Vaccine (4CMenB) in Children and Adolescents: A Systematic Review and Meta-Analysis," *The Lancet Infectious Diseases* 18, no. 4 (2018): 461–472.

31. Taylor, Luke E., Amy L. Swerdfeger, and Guy D. Eslick, "Vaccines Are Not Associated with Autism: An Evidence-Based Meta-Analysis of Case-Control and Cohort Studies," *Vaccine* 32, no. 29 (2014): 3623–3629.

Chapter 8

1. Wood, Bernard, "Evolution: Origin(s) of Modern Humans," *Current Biology* 27, no. 15 (2017): R767–R769.

2. Tattersall, Ian, *Masters of the Planet: The Search for Our Human Origins.* New York: Palgrave Macmillan, 2012; Harris, Eugene E., *Ancestors in Our Genome: The New Science of Human Evolution.* New York: Oxford University Press, 2015.

3. Marks, Jonathan, *Tales of the Ex-Apes: How We Think About Human Evolution.* Oakland: University of California Press, 2015, pp. 107–113.

4. Wood, "Evolution"; Hublin, Jean-Jacques, Abdelouahed Ben-Ncer, Shara E. Bailey, Sarah E. Freidline, Simon Neubauer, Matthew M. Skinner, Inga Bergmann, et al., "New Fossils from Jebel Irhoud, Morocco and the Pan-African Origin of Homo Sapiens," *Nature* 546, no. 7657 (2017): 289–292; Richter, Daniel, Rainer Grün, Renaud Joannes-Boyau, Teresa E. Steele, Fethi Amani, Mathieu Ru., Paul Fernandes, et al., "The Age of the Hominin Fossils from Jebel Irhoud, Morocco, and the Origins of the Middle Stone Age," *Nature* 546, no. 7657 (2017): 293–296.

5. Krings, Matthias, Anne Stone, Ralf W. Schmitz, Heike Krainitzki, Mark Stoneking, and Svante Pääbo, "Neandertal DNA Sequences and the Origin of Modern Humans," *Cell* 90, no. 1 (1997): 19–30.

6. Schmitz, Ralf W., David Serre, Georges Bonani, Susanne Feine, Felix Hillgruber, Heike Krainitzki, Svante Pääbo, and Fred H. Smith, "The Neandertal Type Site Revisited: Interdisciplinary Investigations of Skeletal Remains from the Neander Valley, Germany," *Proceedings of the National Academy of Sciences* 99, no. 20 (2002): 13342–13347.

7. Green, Richard E., Johannes Krause, Adrian W. Briggs, Tomislav Maricic, Udo Stenzel, Martin Kircher, Nick Patterson, et al., "A Draft Sequence of the Neandertal Genome," *Science* 328, no. 5979 (2010): 710–722.

8. Krause, Johannes, Qiaomei Fu, Jeffrey M. Good, Bence Viola, Michael V. Shunkov, Anatoli P. Derevianko, and Svante Pääbo, "The Complete Mitochondrial DNA Genome of an Unknown Hominin from Southern Siberia," *Nature* 464, no.7290 (2010): 894–897.

9. Reich, David, Richard E. Green, Martin Kircher, Johannes Krause, Nick Patterson, Eric Y. Durand, Bence Viola, et al., "Genetic History of an Archaic Hominin Group from Denisova Cave in Siberia," *Nature* 468, no. 7327 (2010): 1053–1060.

10. Prüfer, Kay, Fernando Racimo, Nick Patterson, Flora Jay, Sriram Sankararaman, Susanna Sawyer, Anja Heinze, et al., "The Complete Genome Sequence of a Neanderthal from the Altai Mountains," *Nature* 505, no. 7481 (2014): 43–49; Svante Pääbo, "The Diverse Origins of the Human Gene Pool," *Nature Reviews Genetics* 16, no. 6 (2015): 313; see also Racimo, Fernando, Sriram Sankararaman, Rasmus Nielsen, and Emilia Huerta-Sanchez, "Evidence for Archaic Adaptive Introgression in Humans," *Nature Reviews Genetics* 16, no. 6 (2015): 359–371; Racimo, Fernando, Davide Marnetto, and Emilia Huerta-Sanchez, "Signatures of Archaic Adaptive Introgression in Present-Day Human Populations," *Molecular Biology and Evolution* 34, no. 2 (2017): 296–317.

11. Cleland, Carol E., "Methodological and Epistemic Differences between Historical Science and Experimental Science," *Philosophy of Science* 69, no. 3 (2002): 474–496; Cleland, Carol E., "Prediction and Explanation

in Historical Natural Science," *British Journal for the Philosophy of Science* 62, no. 3 (2011): 551–582; Forber, Patrick, and Eric Griffith, "Historical Reconstruction: Gaining Epistemic Access to the Deep Past," *Philosophy and Theory in Biology* 3, no. e203 (2011). doi: 10:3998/ptb.6959004.0003.003.

12. Hawks, John, "Significance of Neandertal and Denisovan Genomes in Human Evolution," *Annual Review of Anthropology* 42 (2013): 433–449.

13. Marks, *Tales of the Ex-Apes*, p. 114.

14. Yunis, Jorge J., and Ora Prakash, "The Origin of Man: A Chromosomal Pictorial Legacy," *Science* 215, no. 4539 (1982): 1525–1530.

15. https://answersingenesis.org/what-is-science/science-of-uncertainty/ Accessed November 9, 2018.

16. Swinburne, Richard, *The Evolution of the Soul* revised edition. Oxford: Oxford University Press, 1997, p. 1.

17. See, for example, Stephen Jay Gould's praise of Pope John Paul II's claim that evolution is a fact and not in conflict with Christianity (Gould, Stephen J., *Rocks of Ages: Science and Religion in the Fullness of Life*. New York: Ballantine Publishing Group, 1999, pp. 75–82). Also, see Plantinga, Alvin, *Where the Conflict Really Lies: Science, Religion, and Naturalism*. Oxford: Oxford University Press, 2011.

18. McCain, Kevin, and Kostas Kampourakis, "Which Question Do Polls About Evolution and Belief Really Ask, and Why Does It Matter?" *Public Understanding of Science*, 27, no. 1 (2018): 2–10; Kampourakis, Kostas, *Understanding Evolution*. Cambridge: Cambridge University Press, 2014.

Chapter 9

1. Lander, Eric S., "Cutting the Gordian Helix: Regulating Genomic Testing in the Era of Precision Medicine," *New England Journal of Medicine* 372, no. 13 (2015): 1185–1186.

2. Ibid., p. 1185.

3. Kruglyak, Leonid, "The Road to Genome-Wide Association Studies," *Nature Reviews Genetics* 9, no. 4 (2008): 314.

4. Sud, Amit, Ben Kinnersley, and Richard S. Houlston, "Genome-Wide Association Studies of Cancer: Current Insights and Future Perspectives," *Nature Reviews Cancer* 17, no. 11 (2017): 692.

5. Snyder, Michael, *Genomics and Personalized Medicine: What Everyone Needs to Know*. Oxford: Oxford University Press, 2016, pp. 13–18

6. https://www.ispot.tv/ad/7qoF/23andme. Accessed June 15, 2018.

7. Annas, George J., and Sherman Elias, "23andMe and the FDA," *New England Journal of Medicine* 370, no. 11 (2014): 985–988.

8. https://www.fda.gov/NewsEvents/Newsroom/PressAnnouncements/ucm551185.htm. Accessed June 15, 2018.

9. https://www.23andme.com/. Accessed June 15, 2018.

10. https://www.futuragenetics.com/en/. Accessed May 31, 2018.

11. https://www.pathway.com. Accessed May 31, 2018.

12. Fadista, João, Alisa K. Manning, Jose C. Florez, and Leif Groop, "The (In)famous GWAS P-Value Threshold Revisited and Updated for Low-Frequency Variants," *European Journal of Human Genetics* 24, no. 8 (2016): 1202.

13. Burke, Wylie, "Genetic Tests: Clinical Validity and Clinical Utility," *Current Protocols in Human Genetics* 81 (2014): 9–15; Evans, Barbara J., Wylie Burke, and Gail P. Jarvik, "The FDA and Genomic Tests: Getting Regulation Right," *New England Journal of Medicine* 372, no. 23 (2015): 2258–2264; Relling, Mary V., and William E. Evans, "Pharmacogenomics in the Clinic," *Nature* 526, no. 7573 (2015): 343.m

14. Annas, George J., and Sherman Elias, *Genomic Messages: How the Evolving Science of Genetics Affects Our Health, Families, and Future*. San Francisco: HarperOne, 2015, p. 249.

15. https://www.fda.gov/NewsEvents/Newsroom/PressAnnouncements/ucm599560.htm Accessed June 11, 2019.

16. Gill, Jennifer, Adam J. Obley, and Vinay Prasad, "Direct-to-Consumer Genetic Testing: The Implications of the US FDA's First Marketing Authorization for BRCA Mutation Testing," *JAMA* 319, no. 23 (2018): 2377–2378.

17. Tandy-Connor, Stephany, Jenna Guiltinan, Kate Krempely, Holly LaDuca, Patrick Reineke, Stephanie Gutierrez, Phillip Gray, and Brigette Tippin

Davis, "False-Positive Results Released by Direct-to-Consumer Genetic Tests Highlight the Importance of Clinical Confirmation Testing for Appropriate Patient Care," *Genetics in Medicine* 20 (2018): 1515–1521.

18. American Cancer Society. *Cancer Facts and Figures 2016*. Atlanta, GA: American Cancer Society, 2016, p. 11

19. Kurian, Allison W., Emily E. Hare, Meredith A. Mills, Kerry E. Kingham, Lisa McPherson, Alice S. Whittemore, Valerie McGuire, et al., "Clinical Evaluation of a Multiple-Gene Sequencing Panel for Hereditary Cancer Risk Assessment," *Journal of Clinical Oncology* 32, no. 19 (2014): 2001–2009.

20. Additional problems could be caused by mosaicism and chimerism, discussed in Chapter 10.

21. Bloss, Cinnamon S., Nicholas J. Schork, and Eric J. Topol, "Effect of Direct-to-Consumer Genomewide Profiling to Assess Disease Risk," *New England Journal of Medicine* 364, no. 6 (2011): 524–534.

22. Hudson, Thomas J., Warwick Anderson, Axel Artez, Anna D. Barker, Cindy Bell, Rosa R. Bernabe, et al., "International Network of Cancer Genome Projects," *Nature* 464, no. 7291 (2010): 993–998.

23. Brennan, Paul, and Christopher P. Wild, "Genomics of Cancer and a New Era for Cancer Prevention," *PLoS Genetics* 11, no. 11 (2015): e1005522.

24. Annas and Elias, *Genomic Messages*, p. 2.

25. https://www.technologyreview.com/s/610233/2017-was-the-year-consumer-dna-testing-blew-up/. Accessed June 4, 2018.

Chapter 10

1. Thompson, William C., "Forensic DNA Evidence: The Myth of Infallibility," in *Genetic Explanations: Sense and Nonsense*, S. Krimsky and J. Gruber (eds.). Boston, MA: Harvard University Press, 2013, pp. 227–255, at p. 240.

2. It should be noted that whether databases *should* include the DNA of people convicted for some criminal offense or of everyone living in a particular area or state is debatable. There are many ethical and civil liberty issues to consider, and the decision is far from simple and straightforward. In this chapter we refrain from entering this important discussion.

3. Lynch, Michael, Simon A. Cole, Ruth McNally, and Kathleen Jordan, *Truth Machine: The Contentious History of DNA Fingerprinting.* Chicago: University of Chicago Press, 2008, pp. 1–3.

4. https://www.innocenceproject.org/. Accessed June 7, 2018.

5. https://www.newscientist.com/article/2167554-serial-killer-suspect-identified-using-dna-family-tree-website/; https://www.scientificamerican.com/article/the-golden-state-killer-case-was-cracked-with-a-genealogy-web-site1/. Both accessed June 7, 2018.

6. Lynch et al., *Truth Machine*, p. ix.

7. Ley, Barbara L., Natalie Jankowski, and Paul R. Brewer, "Investigating CSI: Portrayals of DNA Testing on a Forensic Crime Show and Their Potential Effects," *Public Understanding of Science* 21, no. 1 (2012): 51–67.

8. Jeffreys, Alec J., Victoria Wilson, and Swee Lay Thein, "Individual-Specific 'Fingerprints' of Human DNA," *Nature* 316, no. 6023 (1985): 76–79.

9. For the various methods used, see Jobling, Mark A., and Peter Gill, "Encoded Evidence: DNA in forensic analysis," *Nature Reviews Genetics* 5, no. 10 (2004): 739–751; Lynch et al., *Truth Machine,* pp. 24–38; Krimsky, Sheldon, and Tania Simoncelli, *Genetic Justice: DNA Data Banks, Criminal Investigations, and Civil Liberties.* New York: Columbia University Press, 2011, pp. 21–22; Kayser, Manfred, and Peter De Knijff, "Improving Human Forensics Through Advances in Genetics, Genomics and Molecular Biology," *Nature Reviews Genetics* 12, no. 3 (2011): 179–192.

10. Krimsky and Simoncelli, *Genetic Justice,* pp. 21–22, 276–277.

11. Thompson, "Forensic DNA Evidence."

12. Jobling and Gill, "Encoded Evidence."

13. Gigerenzer, Gerd. *Calculated Risks: How to Know When Numbers Deceive You.* New York: Simon and Schuster, 2002, pp. 6–7.

14. Lupski, James R., "Genome Mosaicism: One Human, Multiple Genomes," *Science* 341, no. 6144 (2013): 358–359.

15. Wolinsky, Howard, "A Mythical Beast: Increased Attention Highlights the Hidden Wonders of Chimeras," *EMBO Reports* 8, no. 3 (2007): 212–214.

16. Dror, Itiel E., "Biases in Forensic Experts," *Science* 360, no. 6386 (2018): 243.

Chapter 11

1. Rovelli, Carlo, "Science Is Not About Certainty," in *The Universe: Leading Scientists Explore the Origin, Mysteries, and Future of the Cosmos*, John Brockman (ed.). New York: Harper Perennial, 2014, pp. 214–228, at p. 221.

2. Byers, William, *The Blind Spot: Science and the Crisis of Uncertainty*. Princeton, NJ: Princeton University Press, 2011, p. 38.

3. Mercier, Dan, and Hugo Sperber, *The Enigma of Reason*. Cambridge, MA: Harvard University Press, 2017, p. 318.

4. Planck, Max, *Scientific Autobiography and Other Papers*. New York: Citadel Press, 1968, pp. 33–34.

5. Oreskes, Naomi, and Erik M. Conway, *Merchants of Doubt: How a Handful of Scientists Obscured the Truth on Issues from Tobacco Smoke to Global Warming*. New York: Bloomsbury, 2011, p. 34.

6. Potochnik, Angela, *Idealization and the Aims of Science*. Chicago: Chicago University Press, 2017, chapter 1.

7. Aad, G., Abbott, B., Abdallah, J., Abdinov, O., Aben, R., Abolins, M., et al., "Combined Measurement of the Higgs Boson Mass in pp Collisions at $\sqrt{s}$ = 7 and 8 TeV with the ATLAS and CMS Experiments," *Physical Review Letters*, 114 (2015).

8. https://www.economist.com/news/science-and-technology/21710792-scientific-publications-are-getting-more-and-more-names-attached-them-why. Accessed May 1, 2018.

9. Olesko, Kathryn M., "That Science Has Been Largely a Solitary Enterprise," in *Newton's Apple and Other Myths About Science*, Ronald L. Numbers and Kostas Kampourakis (eds.). Cambridge: Harvard University Press, 2015, pp. 202–209.

10. Kampourakis, Kostas, "That Mendel Was a Lonely Pioneer of Genetics, Being Ahead of His Time," in *Newton's Apple and Other Myths About Science*,

Ronald L. Numbers and Kostas Kampourakis (eds.). Cambridge: Harvard University Press, 2015, pp. 129–138.

11. Thurs, Daniel P., "That the Scientific Method Accurately Reflects What Scientists Actually Do," in *Newton's Apple and Other Myths About Science*, Ronald L. Numbers and Kostas Kampourakis (eds.). Cambridge: Harvard University Press, 2015, pp. 210–218, at p. 210.

12. Harris, William, "How the Scientific Method Works," HowStuffWorks. https://science.howstuffworks.com/innovation/scientific-experiments/scientific-method6.htm. Accessed May 2, 2018. .

13. Kuhn, Thomas S., *The Structure of Scientific Revolutions: 50th Anniversary Edition*. Chicago: University of Chicago Press, 2012, p. 174.

14. Kuhn, *The Structure of Scientific Revolutions*, p. 116.

15. Crupi, Vincenzo, "Confirmation," in *The Stanford Encyclopedia of Philosophy* (Winter 2016 edition), Edward N. Zalta (ed.). https://plato.stanford.edu/archives/win2016/entries/confirmation/. Accessed December 20, 2018.

16. Kuhn, *The Structure of Scientific Revolutions*, p. 78.

17. Planck, Max, *Scientific Autobiography and Other Papers*. New York: Citadel Press, 1968, pp. 33–34.

18. Einstein, Albert, *Ideas and Opinions*. New York: Crown, 1954, p. 290.

19. Mercier, Hugo, and Dan Sperber. *The Enigma of Reason*. Cambridge: Harvard University Press, 2017, p. 320.

20. *New York Times* article quoted in Lenz, Gerald E., "Kurt Gödel, Mathematician and Logician," in *The Search for Certainty: A Journey Through the History of Mathematics from 1800–2000,* Frank J. Swetz (ed.). New York: Dover, 2012, pp. 133–135, at p. 133.

21. Ibid., p. 134.

22. Byers, *The Blind Spot*, p. 29.

23. Giaquinto, Marcus, *The Search for Certainty: A Philosophical Account of Foundations of Mathematics*. Oxford: Clarendon Press, 2002, p. 222.

24. Ibid., p. 228.

25. Heisenberg, Werner, "The Physical Content of Quantum Kinematics and Mechanics," in *Quantum Theory and Measurement*, John Wheeler and

Wojciech Zurek (eds.). Princeton, NJ: Princeton University Press, 1983, pp. 62–84, at p. 64.

26. Rovelli, "Science Is Not About Certainty," p. 223.

Chapter 12

1. Kuhn, Thomas S., *The Structure of Scientific Revolutions: 50th Anniversary Edition*. Chicago: University of Chicago Press, 2012, p. 18.

2. Strevens, Michael, "No Understanding Without Explanation," *Studies in the History and Philosophy of Science* 44 (2013): 510–515.

3. Trout, J. D., *Wondrous Truths: The Improbable Triumph of Modern Science*. New York: Oxford University Press, 2016, p. 3.

4. de Regt, Henk W., *Understanding Scientific Understanding*. New York: Oxford University Press, 2017, p. 34.

5. Giere, Ronald N., *Scientific Perspectivism*. Chicago: University of Chicago Press, 2010, p. 63.

6. Potochnik, Angela., *Idealization and the Aims of Science*. Chicago: Chicago University Press, 2017, p. 23.

7. There is debate among philosophers of science concerning whether there can be genuine explanations that are not causal. We do not take a stand on this debate. Rather, we focus only on causal explanations for two reasons: there is no debate concerning whether causal explanations are genuine explanations, and causal explanations are most commonly employed for understanding natural phenomena.

8. See the following helpful *Ted-Ed* video for more on this distinction: Anticole, Matt, "What's the Difference Between a Scientific Law and Theory?" https://www.youtube.com/watch?v=GyN2RhbhiEU. Accessed May 15, 2018.

9. Cartwright, Nancy, *How the Laws of Physics Lie*. Oxford: Oxford University Press, 1983.

10. Kuhn, *The Structure of Scientific Revolutions*, pp. 145–146.

11. For more examples, see McCain, Kevin. *The Nature of Scientific Knowledge: An Explanatory Approach*. Dordrecht: Springer, 2016, chapter 11; and Trout, *Wondrous Truths*.

12. Glymour, Clark, "Explanation and Realism," in *Scientific Realism*, Jarrett Leplin (ed.). Berkeley: University of California Press, 1984, pp. 173–192, at p. 173.

13. Trout, *Wondrous Truths*.

14. de Regt, Henk W., *Understanding Scientific Understanding*, p. 35.

15. Potochnik, *Idealization and the Aims of Science*, p. 23.

16. Kuhn, *The Structure of Scientific Revolutions*, pp. 77–78.

17. Stanford, P. Kyle, *Exceeding Our Grasp: Science, History, and the Problem of Unconceived Alternatives*. Oxford: Oxford University Press, 2006, p. 18.

Chapter 13

1. Hawking, Stephen, and Penrose, Roger, *The Nature of Space and Time*. Princeton, NJ: Princeton University Press, 1996, p. 26.

2. Kampourakis, Kostas, *Making Sense of Genes*. Cambridge: Cambridge University Press, 2017, p. 6.

3. Code, Lorraine, "The Tyranny of Certainty," *Symposium* 21 (2017): 206–218.

4. For helpful discussion of Hempel's view of explanation and prediction, see Okasha, Samir, *Philosophy of Science: A Very Short Introduction*. Oxford: Oxford University Press, 2002, pp. 41–48.

5. Ibid., chapter 3.

6. Douglas, Heather E., "Reintroducing Prediction to Explanation," *Philosophy of Science* 76 (2009): 444–463.

7. Potochnik, Angela, *Idealization and the Aims of Science*. Chicago: University of Chicago Press, 2017, p. 106.

8. McGowan, Kevin J., "White Crows?" http://www.birds.cornell.edu/crows/whitecrows.htm. Accessed May 31, 2018.

9. Cartwright, Nancy. *How the Laws of Physics Lie*. Oxford: Oxford University Press, 1983, chapter 3.

10. Giere, Ronald N. *Science Without Laws*. Chicago: University of Chicago Press, 1999, chapter 5.

11. Hoefer, Carl, "Causal Determinism," in *The Stanford Encyclopedia of Philosophy* (Spring 2016 Edition), Edward N. Zalta (ed.). https://plato.stanford.edu/archives/spr2016/entries/determinism-causal/. Accessed May 31, 2018.

12. Penrose, Roger, *Reality: A Complete Guide to the Laws of the Universe*. London: Jonathan Cape, 2004, p. 806.

13. Isidore, Chris, "These Are Your Odds of Winning Powerball or Mega Millions," CNNMoney. http://money.cnn.com/2018/01/04/news/powerball-mega-millions-odds/index.html. Accessed April 25, 2018.

14. Kampourakis, Kostas, *Turning Points: How Critical Events Have Driven Human Evolution, Life and Development*. Amherst NY: Prometheus Books, 2018.

Chapter 14

1. Schrödinger, Erwin, *Nature and the Greeks*. Cambridge: Cambridge University Press, 1954/1996, p. 90.

2. Bridgman, P. W., *Reflections of a Physicist*. New York: Philosophical Library, 1950, p. 72.

3. De Regt, Henk, *Understanding Scientific Understanding*. Oxford: Oxford University Press, 2017, p. 2.

4. Potochnik, Angela, *Idealization and the Aims of Science*. Chicago: University of Chicago Press, 2017; Elgin, Catherine Z., *True Enough*. Cambridge, MA: MIT Press, 2017.

5. Trout, J. D., *Wondrous Truths: The Improbable Triumph of Modern Science*. New York: Oxford University Press, 2016, chapter 2.

6. Elgin, *True Enough,* p. 45.

7. Ibid., p. 46.

8. Ibid.

9. Strevens, Michael, "How Idealizations Provide Understanding," in *Explaining Understanding: New Perspectives from Epistemology and Philosophy of Science*, Stephen R. Grimm, Christoph Baumberger, and Sabine Ammon (eds.). New York: Routledge, 2017, pp. 37–49, at p. 40.

10. Grimm, Stephen R., "Understanding as an Intellectual Virtue," in *The Routledge Companion to Virtue Epistemology*, Heather Battaly (ed.). New York: Routledge, forthcoming, p. 2. http://www.stephenrgrimm.com/new-page/. Accessed June 13, 2018.

11. Dellsén, Finnur, "Scientific Progress: Knowledge Versus Understanding," *Studies in History and Philosophy of Science* 56 (2016): 72–83, p. 74.

12. De Regt, *Understanding Scientific Understanding*, p. 23.

13. Dellsén, "Scientific Progress," p. 74.

14. Laudan, Larry, "A Confutation of Convergent Realism," *Philosophy of Science* 48 (1981): 19–49.

15. See Psillos, Stathis, *Scientific Realism: How Science Tracks Truth*. New York: Routledge, 1999.

16. Stannard, Russell, *Relativity: A Very Short Introduction*. Oxford: Oxford University Press, 2008, p. 79.

17. Elgin, *True Enough,* p. 60.

18. de Regt, *Understanding Scientific Understanding*, pp. 129–137.

19. Elgin, *True Enough,* p. 59.

Chapter 15

1. Rovelli, Carlo, "The Uselessness of Certainty," 2011: What Scientific Concept Would Improve Everybody's Cognitive Toolkit? https://www.edge.org/response-detail/10314. Accessed July 17, 2018.

2. Rovelli, Carlo, "Science Is Not About Certainty," in *The Universe: Leading Scientists Explore the Origin, Mysteries, and Future of the Cosmos*, John Brockman (ed.). New York: Harper Perennial, 2014, pp. 214–228, at p. 223.

3. Krauss, Lawrence M., "Uncertainty," 2017: What Scientific Term or Concept Ought to Be More Widely Known? https://www.edge.org/response-detail/27085. Accessed July 19, 2018.

4. Ibid.

5. Rovelli, "The Uselessness of Certainty."

6. Burton, Robert Alan, *On Being Certain: Believing You Are Right Even When You're Not*. New York: St. Martins Griffin, 2008, p. 220.

7. For more on the lack of evidence in support of the claim that solar activity is causing climate change see Harker, David. *Creating Scientific Controversies: Uncertainty and Bias in Science and Society*. Cambridge: Cambridge University Press, 2015, chapter 8.

8. Oreskes, Naomi, and Conway, Erik M., *Merchants of Doubt: How a Handful of Scientists Obscured the Truth on Issues from Tobacco Smoke to Global Warming*. New York: Bloomsbury Press, 2010.

9. It would also be helpful if uncertainties in science were not only made clear but communicated in a standardized way. Fischhoff, Baruch, and Davis, L. Alex, "Communicating Scientific Uncertainty," *PNAS* 111 (2014): 13664–13671 make a good case for a set of protocols to help make communicating uncertainties easier for scientists and beneficial to nonscientists.

10. Elgin, Catherine Z., *True Enough*. Cambridge, MA: MIT Press, 2017, p. 7.

11. Firestein, Stuart, *Failure: Why Science Is So Successful*. New York: Oxford University Press, 2016; Firestein, Stuart. *Ignorance: How It Drives Science*. New York: Oxford University Press, 2012.

12. Firestein, *Failure*, p. 249.

INDEX

Tables and figures are indicated by *t* and *f* following the page number.

For the benefit of digital users, indexed terms that span two pages (e.g., 52–53) may, on occasion, appear on only one of those pages.